Martial Gervais ODEN BELLA

Formulation of bioplastic from two biopolymers

Martial Gervais ODEN BELLA

Formulation of bioplastic from two biopolymers

(chitosan and cassava starch) and the setting up of an industrial manufacturing unit

ScienciaScripts

Imprint

Any brand names and product names mentioned in this book are subject to trademark, brand or patent protection and are trademarks or registered trademarks of their respective holders. The use of brand names, product names, common names, trade names, product descriptions etc. even without a particular marking in this work is in no way to be construed to mean that such names may be regarded as unrestricted in respect of trademark and brand protection legislation and could thus be used by anyone.

Cover image: www.ingimage.com

This book is a translation from the original published under ISBN 978-620-2-54707-9.

Publisher:
Sciencia Scripts
is a trademark of
Dodo Books Indian Ocean Ltd. and OmniScriptum S.R.L publishing group

120 High Road, East Finchley, London, N2 9ED, United Kingdom
Str. Armeneasca 28/1, office 1, Chisinau MD-2012, Republic of Moldova, Europe
Managing Directors: Ieva Konstantinova, Victoria Ursu
info@omniscriptum.com

Printed at: see last page
ISBN: 978-620-3-66072-2

Copyright © Martial Gervais ODEN BELLA
Copyright © 2021 Dodo Books Indian Ocean Ltd. and OmniScriptum S.R.L publishing group

FOREWORD

The Ecole Nationale Superieure des Sciences Agro-industrielles (ENSAI.) was created from the former ENSIAAC by order N° 010 CAB/PR on the university reform of January 19, 1993. At its opening, its mission was to train engineers specialized in the agricultural and food industries (IAA). Subsequently, its ambitions were expanded to include the training of young Cameroonian executives in the field of Industrial Maintenance and Production (MIP) from 2000, then in the field of Industrial Chemistry and Environmental Engineering (CIGE) from the 2008/2009 academic year. With the advent of the Bachelor-Master-Doctorate system (LMD), ENSAI has again expanded the training profile of young executives in various fields. In addition to the training cycle of Engineers, ENSAI also prepares students to the diplomas of :

- Professional Master in Quality Control and Management (CGQ), Applied Nutrition (NA), Maintenance of Refrigeration and Thermal Systems (MAINGEFT), Food Science Nutrition (SAN);
- Master's degrees and PhDs in the following departments: Process Engineering (GPI), Applied Chemistry (CA), Electrical Engineering, Power and Automation (GEEA).

During their three-year training, the engineering students of ENSAI carry out each year an internship which, on the academic level, represents a teaching unit (UE) in its own right. The one of the level 3 is qualified as an engineering internship or end-of-study project internship whose main objective is to familiarize the students with a corporate culture, with the challenges of problem solving, that is to say, to facilitate their professional insertion. (4 to 6 months).

The end-of-studies project is not only finalized by a thesis, but also sanctioned by a public defense in front of a jury composed of teachers and possibly industrialists. This project allows the student to deepen his knowledge and techniques and to improve his skills. It is with this in mind that we did an internship at SGB from June 10 to September 30, 2019 (about 4 months). The work focused on the **"Formulation of bioplastic from two biopolymer: chitosan / cassava starch and setting up an industrial manufacturing unit.**

1

SUMMARY

The good mechanical properties and low cost of petrochemical plastics have led to the intensive use of synthetic plastics of petrochemical origin in recent years. Unfortunately, the management of the waste generated by their use is a thorny issue. Several ways are used to remedy this problem : recycling of synthetic polymers, landfilling and/or incineration of the latter within the landfills, implementation of biodegradable polymers of substitution. Our work is focused on the latter way, which protects the environment. The aim is to formulate a biodegradable plastic film by combining chitosan and cassava starch biopolymers, the latter being able to reduce the production cost of the material and improve the properties of the former, and to propose a technical and financial feasibility study for a plastic film manufacturing unit on an industrial scale. Several studies have been made on chitin, the most abundant polysaccharide on Earth after cellulose. Chitosan comes from the deacetylation of chitin contained in the shells of arthropods (crabs, shrimps...) and has film-forming, antibacterial and antifungal properties. In order to achieve our objectives, we have extracted starch and chitosan biopolymer from bitter cassava and shrimp shell waste respectively. Chitin coexists with mineral substances and proteins in shrimp shells. Its extraction was done by a chemical method namely demineralization with 1M hydrochloric acid, deproteination with 1.25M NaOH, then deacetylation with 50% NaOH. The next step was to find the proportion that gives a better property of the plastic film and the method of drying to avoid the attack of bacteria and molds. This resulted in a **starch/chitosan** ratio **of 75:25** and a combined drying in a $300C$ oven **for 12 hours at room temperature for 24 hours.** For an annual turnover estimated at **87 360 000F** CFA the technical study of the project allowed us to dimension the production line and to choose the necessary production equipment. Economically, for an investment of **153 399 362 CFA francs in the first year, the net present value is positive,** the payback period is **3 years 11 months** and the break-even point is **2 years 5 months.** These criteria of appreciation of the project have made it possible to conclude that the project is profitable and viable.

Keywords: cassava starch, chitosan, demineralisation, deproteination, desacetylation, plastic film.

ABSTRACT

The good mechanical properties and the low cost of the petrochemical plastics generate the intensive use of synthetic plastics of petrochemical origin during these last years. Unfortunately, the management of waste generated by their use is a thorny problem. Several ways are used to solve this problem. The synthetic recycling of polymers, the burying and/or incineration of the latter

2

within the discharges. The placement biodegradable substitution polymers. This work concerned the environmental protection by the mention way. It was then a question for to formulate a biodegradable plastic from two biopolymers such chitosan and cassava starch. The last which can reduce the production cost of material and improve the properties of the first and of proposing a technical and financial feasibility study of a manufacturing unit of plastic films on an industrial scale. Several studies were made on chitin and it wasthe most abundant polysaccharide on Earth after cellulose. Chitosan comes from the deacetylation of the chitin contained in the shells of the arthropods (crabs, shrimps ...') and has the property's film forming, antibacteriennes and antifongic. To attain our objectives, there was some discussion about extracting the biopolymers from starch and chitosane starting from bitter cassava and from shrimp waste shells respectively. Chitin coexists with mineral substances and proteins in the shrimps shells.Its extraction was made by a chemical method which consisted of demineralizingwith hydrochloric acid 1M, deproteination by NaOH 1,25M, then a desacetylation by NaOH with 50%.The following stage consisted in finding the proportion which gives a better property of plastic film and the method of drying to avoid the attack of the bacteria and moulds. It arises that the proportions **starch/chitosan of 75/25** and the combine drying method in an oven **30^0C for 12 hours at an ambient temperature for a period of 24 hours**. For an annual turnover estimated at **87 360 000 F CFA**, the technical part of the project studied permitted to choose the necessary production equipment. For investment of **153 399 362F CFA** for the first year. The **Net present value is positive,** recovery time **3 years, 11 months** and death point **2 years, 5months**. These evaluation criteria of the project permitted to conclude on the profitability and viability of the project.

Key words: cassava starch, chitosan, demineralization, deproteination, desacetylation, plastic film.

Table of contents

4

INTRODUCTION

Nowadays, cities in Africa and elsewhere are confronted with the problem of waste management in general and plastic waste in particular. A plastic that has become waste, that is to say, abandoned or destined to be abandoned, is often discarded in our streets, gutters and sometimes in our oceans, leading to a recurrence of floods, especially in precarious areas. Not being biodegradable, plastic waste causes long-lasting pollution. It is the source of significant air pollution when burnt in the streets. Cameroon has few structures for the collection and recycling of plastic packaging, so a problem arises over time: that of the treatment of the said packaging. Indeed, according to a survey conducted by MINEPDED in 2011, 58% of consumers dispose of their used plastic packaging in the streets, holes or bins. These conventional plastic packaging uses therefore pollute the environment quite significantly. This is what led the Ministers of Environment, Nature Protection and Sustainable Development and that of Trade to sign the joint order No. 004 / MINEPDED / MINCOMMERCE Portant regulation of the manufacture, import and marketing of non-biodegradable packaging prohibiting non-biodegradable plastic packaging on 24 October 2012. The Minister of Environment, Protection of Nature and Sustainable Development issued a press release 24 hours before the repression brigade's raid on the field, marked with the seal "Urgent". In the administrative document, the minister no longer speaks of banning non-biodegradable plastic packaging, but rather of banning "plastic packaging with a thickness of less than or equal to 60 microns". MINEPDED explains that to date, biodegradable plastic packaging is not manufactured in Cameroon. "There are economic operators who rather manufacture oxo-degradable plastic packaging which are fragmentable plastics and not biodegradable. These plastics are just as, if not more dangerous than ordinary plastics," he explains. One of the strategies for solving this pollution problem is the complete recycling of the waste generated by oil-based polymer materials. However, the recycling of these materials is limited and consumes a considerable amount of energy. Biological recycling or biodegradation of polymers can then be considered as an alternative to more traditional recycling procedures, which has led researchers to synthesize and use new polymers that can be returned to the biological cycle after use. Therefore, the use of natural biopolymers that are easily biodegradable would solve these problems. In this context, there is currently a renewed interest in the development of biodegradable materials from natural biopolymer (Moon et *al.,* 2000). Among natural biopolymer, starch is considered the most promising raw material for the development of new and more environmentally friendly materials, especially for packaging and disposable material applications (Mutungi et *al.*, 2011). This is due to its low density, renewable nature and complete biodegradability, availability worldwide in different forms and low cost. The main disadvantages of amylaceous materials compared to oil-based polymer materials are their hydrophilic character and their poor mechanical properties which lead to low stability. In order to replace the latter materials by the former, these disadvantages must be overcome and the quality and properties of the plastic film must be improved, hence the mixture with chitosan, due to its film-forming properties, its natural origin, its biocompatibility, its biodegradability or its antibacterial properties. The problem that is highlighted here is that of the need to create and perpetuate a company for the production and marketing of biodegradable plastic packaging, but before reaching this final point, it will first be necessary to carry out a technical feasibility study that will allow the emergence of all the necessary tools for the implementation of this unit. The overall objective of the project is to assess the technical feasibility of producing biodegradable packaging and setting up a production and marketing unit. Specifically, it is about :

-   Extraction and characterization of the extracted biopolymers;
-   Make a formulation and implementation of the plastic film;
-   To carry out a technical and commercial feasibility study of a manufacturing unit on an industrial scale;
-   Make a financial study of the project implementation

The present document is the quintessence of the work done. It is divided into three chapters. The first chapter is reserved for a review of the literature on the concepts addressed, the second for the methodological approach adopted and the different tools used to carry out this work and finally the third chapter presents the main results obtained and discussed. The third chapter presents the main results obtained and discussed. We will end this paper with a conclusion and perspectives on this work.

I.1. Classification of biopolymer

There is a wide variety of biodegradable polymers. (Legros et *al.,* 2011) proposed a classification of biopolymer according to the manufacturing method, origin and nature of the raw material. Figure 2 shows the types of biopolymer according to the manufacturing method. Types 1 to 3 are derived from renewable resources (biosources) and type 4 is derived from non-renewable resources.

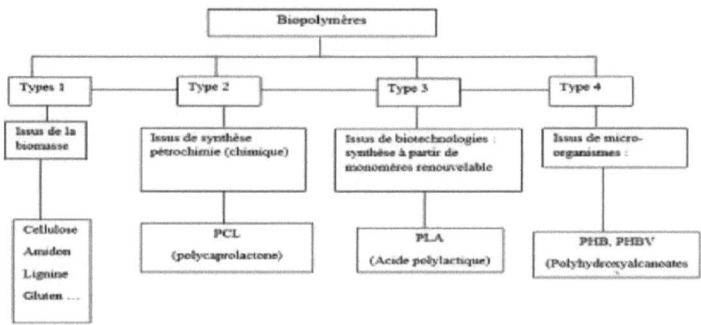

Figure 1: Classification of biopolymer. (legros et al, 2011)

I.2. STARCH-BASED BIOPOLYMERS
I.2.1. primary composition and structure of starch

Starch is a carbohydrate with the gross formula **(C6 H10 O5) n**. It consists of 98-99% of a mixture of two natural polymers: amylose and amylopectin, which are composed of a-D-glucopyranose (or a-D-glucose or anhydroglucose) molecules and are present in a cyclic form. Amylose is a nearly linear macromolecule and amylopectin is a highly branched macromolecule. The other constituents (1-2%) of starch are lipids, proteins, minerals and phosphorus located both on the surface of the starch and in the interior. These constituents, although minor and present in small quantities, are capable of modifying the physico-chemical properties of starch. Starch is the bio-synthetic carbohydrate compound and the main source of energy for human and animal life. It represents an important weight fraction in a large number of agricultural raw materials such as cereals (30-80% of dry matter), legumes (25-50% of dry matter) and tubers (60-90% of dry matter). Its low price makes starch an economically interesting material.

Table 1: Amylose and amylopectin contents (% DM) of natural starches according to Cheftel and Cheftel (1976)

Potential source	Amyloidosis	Amylopectin
Potato	23	77
Cassava	20	80
Ble	20	80
Rice	15 a 35	65 a 85
Sorghum	25	75
But	25	75
But waxy (1)	0	100
Amylomais (2)	77	23
banana	17	83

(1) and (2) obtained by genetic modification

From the point of view of amylose content (which depends on the botanical origin), there are three types of starch: standard starch (20-35% amylose), waxy starch (less than 15% amylose) and mutant or amyloid starch (more than 40% amylose). Some species with so-called mutant genotypes have very low or very high amylose contents. This is the case of waxy maize (less than 1% amylose) and amylomais (nearly 80% amylose).

I.2.2. Amyloidosis

Amylose is an essentially linear macromolecule consisting of D-anhydroglucopyranose units linked mainly by a-linkages (1-4). There are a few branching points (a (1-6) linkages). The higher the molecular weight, the greater the number of branches. The two ends have different functions. At the C4 position is the non-reducing end and at the C1 position the reducing end due to a hemiacetalic function of the terminal secondary alcohol. Native amylose contains 500 to 6000 glucosyl units depending on the botanical origin, distributed in several chains with an average degree of polymerization (DP) of about 500 and a weight average molecular weight of 105 to 106 (Banks and Greenwood, 1975), which is ten times greater than that of conventional synthetic polymers.

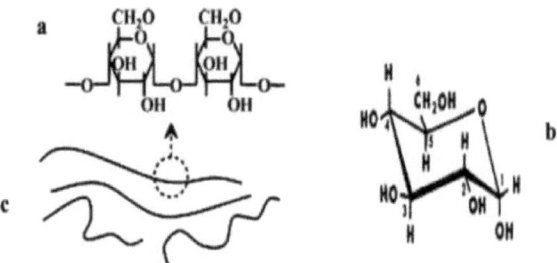

Figure 2: Chemical structures of : (a) amylose, (b) d-glucopyranose. schematic representation of amylose (c) (liu et al., 2009).

I.2.3. Amylopectin

Amylopectin is branched amylose (Figure 4). It consists of 10,000 to 100,000 repeating units of glucose. In addition to the a (1 ^ 4) linkages of amylose, 4 to 5 % branching points are present in amylopectin through acetal (1 ^ 6) linkages. Its degree of polymerization and molar mass are in the ranges 960015900 and 107-109 respectively, depending on the botanical origin of the starch. The high molar mass of amylopectin and its branched structure reduce the mobility of polymeric chains.

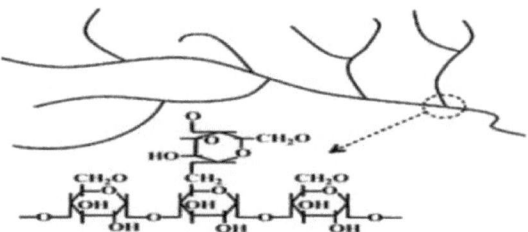

Figure 3: Chemical structure and schematic representation of amylopectin (liu et al., 2009)

I.2.4. Morphology and fine structure of starch
a) Morphology of starch grains

After extraction and purification, starch is a white powder insoluble in cold water. This powder, depending on the botanical origin, is made up of dense microscopic (= 1.5) starch storage units ranging in size from 1 to 100 pm. These entities are called starch grains and their shape also varies according to their botanical origin.

Observed under polarized light microscopy, starch grains are birefringent and show a Maltese cross whose branches meet at the hilum (starting point of the starch grain growth) (Biot, 1844). The birefringence is positive which implies a radial organization of macromolecular chains within the grain.

 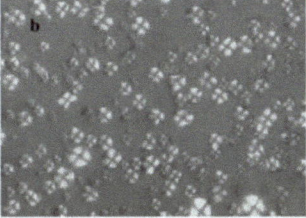

Figure 4: Scanning electron microscopy observations of cassava starch grains
(a) and polarized light microscopy of waxy maize grains (b) (Valbiom, 2011).

b) Hydrothermal behaviour of starch grains

The starch grain, when heated in the presence of an excess of water at a
temperature above 60°C, goes through three successive states: swelling,
gelatinization and solubilization. A physical gel is formed during cooling: this is
retrogradation.

c) Gelatinization

Gelatinization or stacking is an irreversible swelling and partial
solubilization of the starch grain in the presence of excess water and at
temperatures above 60°C. During heating, the starch grains absorb water in the
amorphous zones. This swelling leads to the destructuring of the grain (breaking
of hydrogen bonds in the crystalline zones, disappearance of the Maltese Cross).
As the hydrogen bonds are broken, the amylose diffuses out of the grain and
solubilizes in the medium: this is the starching phase which leads to the formation
of a starch paste. This starch paste is a suspension in which the swollen starch
grains constitute the dispersed phase, and the solubilized macromolecules of the
amylose form the continuous phase.

d) Retrogradation and Freezing

Retrogradation refers to the structural reorganization (or recrystallization) that
takes place during the cooling of a destructured starch suspension. During
cooling, when the concentration of the polymer is sufficient (> 1%), a white
opaque gel is formed: this is gelification. It takes place in two stages characterized
by a statistical ball/double helix transition at the polymer chain segments and then
by crystallization by stacking of the chains (Buleon et *al.,* 1990). Amylose freezes
rapidly (Miles et *al.,* 1985) whereas amylopectin freezes more slowly and is
limited by its branching structure (Ring et *al.,* 1987). Starch gels can be
considered as composite materials with amylose gel as matrix and amylopectin-
rich grain ghosts as reinforcement.

The gels consist of two phases and the composition of each phase depends mainly on the degree of gelatinization and the amylose/amylopectin ratio of the starch grain (Buleon et *al.*, 1990). Melting of recrystallized amylose occurs at about 120°C, whereas retrograde amylopectin crystals melt at low temperatures (Miles et *al.*, 1985). The retrograde crystal structures are of type B regardless of the macromolecule. Retrogradation governs the formation of starch-based films.

The gelification of starch is a phenomenon essentially initiated by a decrease in temperature, which in turn induces a decrease in polymer solubility. The resulting gels are porous structures (Leloup *et al.*, 1990).

I.2.5. Starch applications

An impressive range of applications: stabilizers in soups and frozen foods, tablet coating and paper coating, adhesives for stamps and counterplates, textile dressing, raw material for ethanol production, and even concrete binder, thermoplastic film. (FAO, 2006).

I.3. CHITIN AND CHITOSAN

In 1811, Prof. Henri Braconnot, Director of the biological garden in Nancy (France) isolated a fibrous substance from a certain type of mushroom. Moreover, he observed that this substance is not soluble in aqueous solutions of acids. A decade later, in 1823, the same substance was found in some insects (coleopteran) and was later named chitin (from the Greek word "kitos" which means envelope). In 1859, Prof. C. Rouget subjected chitin to alkaline treatment and observed the different solubilities of chitin. The substance, resulting from the alkaline treatment, could be dissolved in acids. However, only in 1894 this substance was named chitosan by Hoppe-Seyler (Jaouen, 1994). Between 1930 and 1940, these biopolymeres aroused a lot of interest in the Eastern world, mainly for application in the medical field and water purification. Today, we know that chitin and chitosan are abundantly found in nature and are renewable resources (Muzzarelli, 1977).

I.3.1. Chitin

Chitin is a natural polysaccharide widely distributed in nature. In the animal kingdom, chitin is an important structural element of the teguments of certain invertebrates; it exists only in the form of complexes with proteins and minerals. It is present mainly in the shells of molluscs, in the cuticles of insects and in the carapace of crustaceans.

In the plant kingdom, chitin is found in the walls of most fungi and some chlorophyceous algae. In addition to its role in maintaining cell rigidity, it contributes to the control of osmotic pressure. Chitin is also present in some yeasts and bacteria.

Although there are many potential sources of chitin (Table 3), it is mainly produced today from shrimp shells. For a long time, these wastes were not recycled and were simply thrown back into the sea after peeling. The production of chitin makes it possible to valorize the waste of the food industry by avoiding that they are rejected at sea, which generates problems of pollution because the carcasses of arthropods (crustaceans, cephalopods...) are very resistant to biodegradation (Shahidi and Abuzaytoun, 2005). Crustacean shells contain about 30-40% protein, 30-50% calcium carbonate, and 20-30% chitin on a dry weight basis (Johnson and Peniston, 1982).

Chitin comes mainly from the shrimp shell, whose composition in mass for the case of shrimp *Palaemon fabricius* on average is the following:

- 75% water
- 12% protein
- 9% mineral salts
- 4% from China
- Trace of lipids and organic pigments

One kilogram of fresh shells yields about 40 grams of dry chitin. After chemical treatment, the grinding and sieving processes to obtain a homogeneous chitin result in losses of about 40%. The final yield is 2.5%, i.e. 25 grams of chitin per kilogram of carapace. (Valentin DURAND and Thomas VERGINI, 2010).

Table 2: Potential sources of chitin (Tolaimate et al, 2003).

Sources of chitin	Chitin content (%)
Barnacle *(Lepas anatifera)*	7
Marble crab *(Grapsus marmoratus)*	10
Red crab *(Portunus puber)*	10
Spider crab *(Maia squinado)*	16
Lobster *(Homarus vulgaris)*	17
Grasshopper lobster *(Scyllarus arctus)*	25
Lobster *(Palinurus vulgaris)*	32
Crayfish *(Astacus fluviatilis)*	36
Shrimp *(Palaemon jabricius)*	22
Cuttlefish *(Sepia officinalis)*	20
Squid *(Loligo vulgaris)*	40

I.3.2. Process for the extraction of chitin

Many methods have been developed to prepare chitin from shells. In general, they consist in removing mineral elements (demineralisation), proteins (deproteinisation) and colour (bleaching).

a) Demineralization

Acid treatment removes the minerals, which pass into solution as salts. For economic reasons, chloridic acid (HCl) is preferred. The minimum concentration for this step is determined by the chemical equation of the reaction [$CaCO_3 + 2HCl \wedge CaCl_2 + CO_2$ (f) + H_2O] between the major mineral, calcium carbonate, and HCl. In principle, the demineralization is complete when the proportions are strenchiometric. The HCl concentrations encountered are between 0.5 and 11 N and the substrate/solvent ratio between 1:10 and 1:40. Demineralization takes between 15 min and 48 h, from room temperature to 50 °C (Tolaimate et al. 2003; Al Sagheer et al. 2009).No et al. (1989) recommended using a commercial antifoam containing 10% active silicone solution.

b) Deproteinization

A basic treatment allows the removal of proteins by solubilization. The reagents used for this step are strong bases such as potassium hydroxide (KO H). The most common, for economical and technical reasons, is sodium hydroxide (NaOH). The concentrations used are between 0.3 and 2.5 M, in a ratio of between 1:10 and 1:40. The temperature is between 50 and 110 °C and the duration can vary from 1 h to more than 24 h (Tolaimate et al 2003; Al Sagheer et al. 2009).

c) Bleaching

Generally, the bleaching agent used is hydrogen peroxide (H202), with a concentration of 0.1 to 33%, and it can also be mixed with HCl. The treatment time is often very short, in the order of a few minutes (Tolaimate et al. 2003).

I.3.3. Chitosan

Chitosan is obtained by partial N-desacetylation of the chitin molecule. Their chemical structure, shown in Figure 6, results from the linking of N-acetyl-D-glucosamine and D-glucosamine repeat units in p, (1^4). Chitin and chitosan differ in the proportion of acetylated units present in the copolymer, also called the degree of acetylation (note DA). Although the term "chitosan" is usually restricted to any chitin sufficiently N-deacetylated to be soluble in dilute acid, there is no official nomenclature proposing the precise boundary between the two terms. Accordingly, we will refer to any sample with residual acetylation degrees (AD) < 30% as chitosan (Seng, 1988).

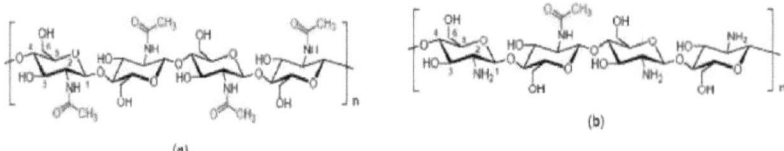

(a)

(b)

Figure 5: Chemical structures of chitin (a) and chitosan (b)

Chitosan has three types of functional groups, an amine group and two hydroxyl groups at the C2, C3 and C6 positions respectively. Due to its free amine groups, this compound has very interesting properties. Chitosan, soluble in weak acids, has a positive charge whereas most polysaccharides are negatively charged (Shahidi et *al.,* 1999). The degree of deacetylation (DD) represents the proportion of amine-Dglucosamine units to the total number of glycoside units. In the case of chitosan, the DD is above 60% for commercial products. This value also determines the solubility limit of this polymer in dilute acid solutions (2< pH <6). The DD is a structural parameter that influences physical-chemical properties such as solubility, bulk charge and reactivity, mechanical properties such as elongation at break, tensile strength and barrier properties. It also influences biological properties (Chatelet et al., 2001) such as biocompatibility, biodegradability, and biostimulant and anticholesthetic activity.

I.3.4. Production of chitosan from chitin

With the present techniques it is possible to achieve a suitable demineralisation and deproteinisation of the shell from the waste material and thus a partially purified chitin. Depending on the requirements, further purification can be carried out to obtain an ultra-pure product free of any protein. Starting from chitin, the degree of acetylation and the molar mass of the polymer can also be adjusted as desired, in order to obtain various chitosans by means of controlled basic hydrolysis. If chitin extraction is nowadays easily achievable, the most delicate step remains deacetylation which requires sufficient substitution of the acetyl groups to obtain chitosan, which can lead to an excessive reduction of the polymer chain length. By adjusting the duration of the alkaline treatment and the temperature rise, it is therefore possible to obtain different chitosans from the same chitin.

A DA of less than 10% is rarely achieved by a simple process, total deacetylation requires several steps. The sample is either washed or dissolved and reprecipitated between two cycles. Deacetylation is usually carried out by treatment with concentrated soda or potash (40-50%) at a temperature > 100 °C, for at least 30 minutes to remove some or all of the acetyl groups from chitin (No and Meyers, 1995). Deacetylation cannot be accomplished with an acidic reagent (Muzzarelli, 1977). There are several essential factors that affect the yield of deacetylation such as temperature, time, concentration of alkali, prior treatments for obtaining chitin, atmosphere (air or nitrogen), amount of chitin to alkali solution, chitin density and particle size. Taking into account these parameters, the ideal objective of deacetylation is to prepare a chitosan which is undegraded and perfectly soluble in dilute acid (acetic, lactic, citric acids,...).

I.3.5. Physicochemical properties of chitin and chitosan

a) Degree of acetylation (DA)

Chitin and chitosan are mainly characterized by their DA and by their molar mass, the determination of which requires the solution of the polymer. The degree of acetylation presents the ratio of the acetyl group to the non-acetyl group. Many methods are proposed to determine the DA on solid samples: IR spectroscopy (Brugnerotto, 2001), elemental analysis (Kasaai et al., 1999), solid NMR (Heux et al., 2000), or on samples in solution: UV determination (Maghami and Roberts,

1988), colloidal titration (Chen et al., 1997), liquid NMR (Yang and Montgomery, 2000).

b) **Solubility**

Chitin is soluble in very few solvents. Its initial molar mass is of the order of 800 000 to 106 g.mol-1 and it is generally highly acetylated. Industrially obtained chitosans have a molar mass of the order of 200 000 g.mol-1 and a DA ranging from 2 to 25%. For chitosan, the molar mass and the distribution of N-acetylated units along the chain are dependent on the deacetylation method used. In addition, a successive sequence of several N-acetyl units gives the polymer a more hydrophobic character, and therefore self-associating properties (thickening and gelifying properties) and modifies its solubility.

c) **Viscosity**

The viscosity of chitosan depends on the degree of deacetylation (DD) of this polymer. The more it is deacetylated, the more free amine groups there are, the more soluble the chitosan is and consequently its viscosity is higher. The viscosity also depends on: the concentration of the polymer (it increases with the concentration), the temperature (it drops when the temperature increases), the molecular weight (the intrinsic viscosity increases with the increase in molecular weight) and finally the pH (the lower it is, the higher the viscosity) (Berth et al., 1998).

I. 3.6. Applications of chitin and chitosan

The particular properties of chitosan are essentially linked to the presence of the amine function carried by carbon 2. They are of two types:

At acid pH, where it is soluble, chitosan carries many positive charges. It is therefore a good flocculating agent and a good coagulating agent. It can interact with negatively charged molecules (e.g. fatty acids, proteins ...) (Fang et *al.,* 2001) and thus form polyanion-polycation complexes. This property opens up a wide range of applications, particularly in dermo-cosmetology: it allows it to interact very intimately with the skin's keratins and thus form a film on the skin's surface.

- At pH (> 6.5), chitosan loses its positive charges, the nitrogen electronic doublet is free. These free doublets and the presence of many oxygen atoms in chitosan allow it to behave as an excellent complexing agent, especially for heavy metals. It is therefore used for water purification, for example for the recovery of metals from industrial effluents.

In the field of cosmetics, the film-forming and cationic properties of chitosan are exploited in many hair and skin care creams or lotions (Lang and Clausen, 1989). It is found, for example, in antifungal, slimming and moisturizing creams...

I.   3.7 Chitin/chitosan films

Chitin and chitosan films are obtained by dissolving them in an acid solution. The moisture content of the polysaccharide must be low to make the film. From the point of view of their mechanical characteristics, chitin films cannot undergo elongation. Chitosan-based films can be stretched only twice. These mechanical properties can be improved by modification of the chitosan (by branching) or by formulation adjustments. For example, chitosan-poly(vinyl alcohol) has better mechanical strength, stability and elongation capacity (Nakano, 2007).

I.4. WAYS OF VALORIZATION OF SHRIMP SHELLS AND HEADS

Co-products are defined as those parts which are not used and which can be recovered during traditional production operations. In the case of shrimps, this means the heads, shells and tails. Marine by-products make up 30 to 60% of the whole product and for several years now they have been attracting the attention of industry for reasons of economic profitability and sustainable development. Indeed, these materials contain many valuable molecules including proteins (Heu *et al.*, 2003), lipids (Dumayet *al.*, 2006), minerals, vitamins, and other bioactive compounds (Kim *et al.*, 2008), beneficial to human and animal health. Figure 7 presents the main ways of valorisation of marine co-products.

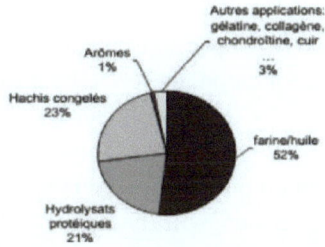

Figure 6: Proportion of the different recovery routes for marine co-products.
(Andrieux, 2004)

I.5.   PRODUCTION TECHNOLOGY FOR PLASTIC FILMS

1.5.1. Film formatting

Two basic technologies, namely dry and wet processes, are used for the preparation of biopolymer films.

1.5.1.1. Dry processes

Dry processes have been mainly applied to the preparation of packaging biomaterials with plasticized starch and protein using conventional processing techniques such as extrusion moulding, compression and injection (Chandra and Rustgi, 1998). Extrusion of biopolymers is the preferred method for high volume production for applications such as packaging. The thermo-compression method is also useful as a processing method because of its simplicity and ability to produce films without solubilization. Thermo-compression is a technology that allows the bonding of heterogeneous materials under the simultaneous action of temperature and pressure without the formation of any liquid phase during the bonding process, all in a variable atmosphere. In these processes, the specific mechanical energy, shear impact, pressure, plasticizer, time and temperature are important parameters in determining the film properties.

Although dry processes require more equipment, they have some major advantages over wet processes: they are closer to industrial implementation and they considerably decrease the solubility of the resulting films by creating a highly cross-linked film network (Rhim and Perry, 2007).

1.5.1.2. Wet processing

There are different methods for making wet laid films. The "*casting*" method or the method of casting and evaporation of the solvent is based on the drying of the film-forming solution. It includes solubilization, casting and drying. The first step is to prepare a film-forming solution by dissolving a biopolymer in a suitable solvent such as water, alcohol, or an organic solvent. The solution is then poured into an anti-adherent mould and evaporated at room temperature or at high temperatures. The actual formation of a cohesive film depends on the nature, type and extent of the interactions of the polymers involved as well as on the film forming conditions such as temperature and drying rate, moisture content, type of solvent, plasticizer concentration and pH. The *casting* method has been commonly used for the preparation of biopolymer-based films (Souza et *al.,* 2011).

1.5.2. Effects of plasticization on polymer properties

The properties of a polymer will change after plasticization. The improvement of the flexibility of a resin by plasticizers is one of the properties that has been researched for a long time. Plasticizers are also used to reduce brittleness, decrease hardness, decrease crystallinity, glass transition temperature and melting temperature. These changes will directly impact the shaping of the polymer and make it easier. The addition of plasticizer to a formulation will also decrease the cost of a plastic product (Mekonnen, et al., 2013).

1.5.3. Parameters influencing lamination

Plasticization of polymers is used for two different purposes: to assist in the shaping of the polymer and/or to modify the final product. In order to achieve the set objective(s) it is possible to vary certain parameters such as the plasticizer and/or the processing.

1.5.4. Choice in the use of plasticizer

It is important to choose a plasticizer with excellent compatibility with the polymer, an effective plasticizing action and good stability (from a thermal, oxidizing, UV point of view...). Other properties may also be of interest when making a selection: biodegradability, non-volatile and non-toxic plasticizer and low migration over time. Water can be a good plasticizer for polysaccharides but processing temperatures used in plasticization can be above 100 °C (Mekonnen, et al., 2013).

Depending on its structure, each polymer will have adapted plasticizers. For starch, the most used plasticizers are generally polyols like glycerol, sorbitol, xylitol, maltitol, ethylene glycol, propylene glycol, etc. For cellulose, which is also a polysaccharide and the most abundant biopolymer on Earth, the plasticizers that are commonly used are: diethylephthalate, dimethyle phthalate, triphenyl phosphate, ethylhexyl adipate, triacetin, glycerol triacetate, triethylene, etc. (Mekonnen, et al., 2013).

A wide range of plasticizers exists, the choice of one of them is essential to obtain a sufficient plasticization to facilitate the implementation or the properties of the polymer to be plasticized.

1.6. Biodegradable starch-based plastic film

In order to manufacture bioplastic more respectful of the environment, many studies have been carried out among these studies we can quote:

Elaboration and characterization of biofilms based on cassava starch reinforced by two- and three-dimensional mineral fillers" Pierre Celestin Belibi, thesis carried out under international co-supervision and submitted to obtain the degree of Doctor /PhD in Physics from the University of Yaounde I, specializing in Materials Science.

> the realization of a plastic film based on corn starch /chitosan, those of (Carole ., 2019) thesis of end of [2nd] cycle, Enset of Douala "Experimental study and design of an industrial process of manufacture of a suit of

> conforming safety for surgeon based on 100% cotton fabric and biodegradable and antibacterial plastic film".

Figure 7: Presentation outfit "NGOUA" source: memoire carole ., 2019.

CHAPTER II: MATERIAL AND METHOD

This part aims to present the tools, materials, different techniques used and the experimental protocols applied for the extraction of starch, chitin and chitosan as well as the formulation of the plastic film. The extraction of chitin is carried out in three steps: 1) depoetization, 2) demineralization and 3) removal of lipids and pigments. Chitin can be converted to chitosan by different processes: homogeneous, heterogeneous or enzymatic alkaline deacetylation, or by the TMC (Thermo-Mechano-Chemical) method. In this work, we proceeded to the homogeneous alkaline deacetylation.

11.1. Material

11.1.1. Documents

4- Survey sheet on the availability of shrimp shells in Douala city (annex 3) ;

4- **Administrative procedures to follow for the creation of a limited liability company (S.A.R.L);**

4- **Taxes or charges imposed.**

11.1.2. Biological material

The first plant material used in this work is bitter cassava tubers (*Manihot Esculenta*). It lends itself well to production especially since it is richer in starch. Cultivated in the Noun department, more precisely in the Njimom district in the Maloure village.

Photo 1: Fresh cassava tubers (a) and washed peels (b)

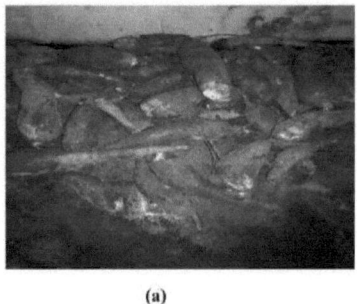 

(a)   (b)

The second biological material is the shells of shrimps collected from the sea.

(a)   (b)

Photo 2: Fresh shells collected (a) and washed (b)

II.1.3. Devices
Table 3: Devices used

N°	Désignations	Rôles	Images
1	**Le thermomètre :** c'est un pyromètre optique (limites extrêmes comprises entre 50°C et 550°C)	Maîtriser la température des solutions et de l'évaporation d'eau pendant la cuisson du mélange ainsi que celle du séchage.	
2	**Le pH-mètre**	Déterminer le pH de l'amidon, chitine et chitosane	
3	**Le four microonde**	Déterminer la teneur en cendres	
4	**Etuve**	sécher les échantillons	

II.   1.4. Reagents
Table 4: Reagents used

N°	Produits	Propriétés physico-chimiques
1	Acide chlorhydrique	Liquide fluide ; acide fort (33%) ; soluble ; densité à 20°C :1.011+/-0.01
2	Hydroxyde de sodium	Cristaux ; Base forte ; minimum 99% de pureté ; la solubilité augmente avec la température; NaOH
3	Peroxyde d'hydrogène	Fluide ; densité : 1,573+-0,001 ; décomposé par de nombreux solvants organiques ; H_2O_2
4	L'acétone	C_3H_6O ; densité 2,88+-0,03 ; T° ébullition 56,05°C ; miscible avec l'eau,
5	Eau distillée	PH=5,4 ; très instable, conductivité électrique quasiment nulle
6	Acide acétique	CH_3COOH Liquide fluide ; acide faible (99,9%) ; soluble ; densité à 20°C ;

II.   1.5. Software
- **The Word software version 2016**: It is the software that allowed the word processing as well as the representation of the manufacturing diagrams.
- **The Excel software version 2016**: It allowed to make a financial evaluation of the project.
- **The Visio 2016 software**: It allowed to represent the organization chart of the company

Archicad Software 2018: The archicad software was used to represent the company's construction plan in 2D. Indeed, it is an architect's software allowing the construction of house plans.

II.2 Methods
II.2.1 Summary of the work

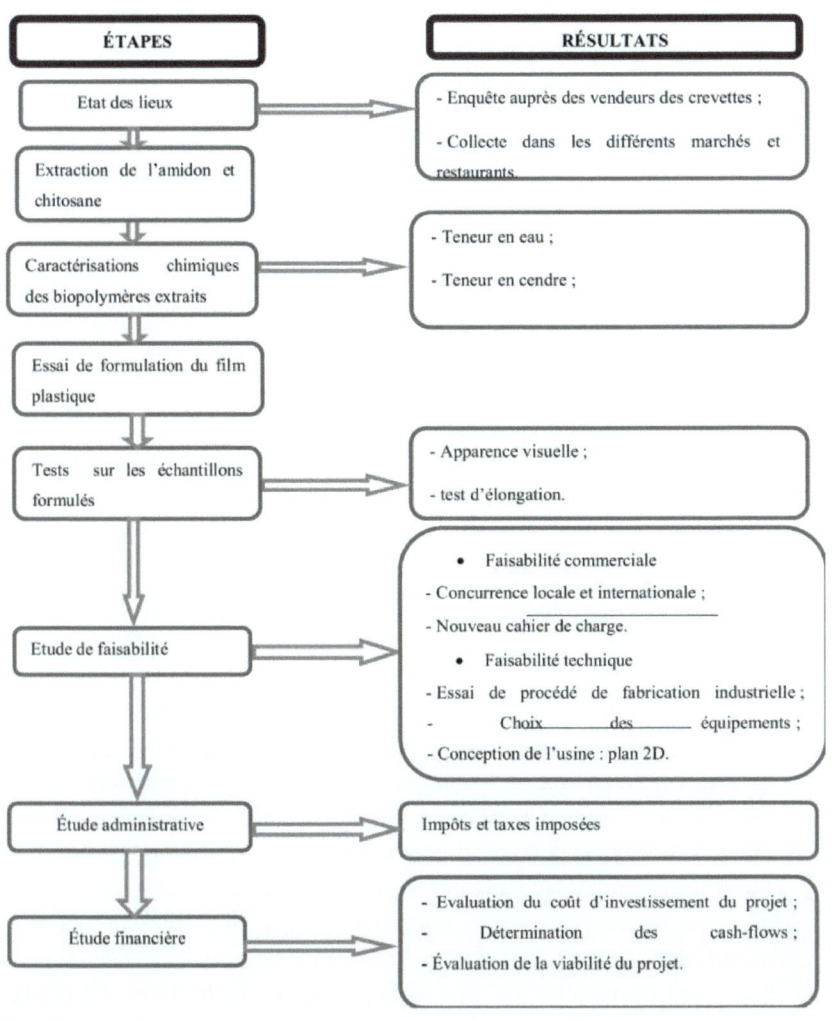

ÉTAPES

- Etat des lieux
- Extraction de l'amidon et chitosane
- Caractérisations chimiques des biopolymères extraits
- Essai de formulation du film plastique
- Tests sur les échantillons formulés
- Etude de faisabilité
- Étude administrative
- Étude financière

RÉSULTATS

- Enquête auprès des vendeurs des crevettes ;
- Collecte dans les différents marchés et restaurants.

- Teneur en eau ;
- Teneur en cendre ;

- Apparence visuelle ;
- test d'élongation.

- • Faisabilité commerciale
- Concurrence locale et internationale ;
- Nouveau cahier de charge.
- • Faisabilité technique
- Essai de procédé de fabrication industrielle ;
- Choix des équipements ;
- Conception de l'usine : plan 2D.

Impôts et taxes imposées

- Evaluation du coût d'investissement du projet ;
- Détermination des cash-flows ;
- Évaluation de la viabilité du projet.

Figure 8 : Schéma synoptique du travail

With the help of a survey form shown in Annex 2, we visited the markets and large restaurants in Douala to get an idea of the type and quantity of shrimps sold and the quantity of shells generated per day.

II.2.3. Starch extraction and chitosan

II.2.3.1. Extraction of cassava starch

Cassava is a root that can stay in the ground for almost two years, but once harvested it deteriorates very quickly. That is why very often part of the harvest is processed (fermented or not flour, tapioca, attieke or starch). Before processing, we selected healthy, ripe, firm, freshly harvested cassava roots in order to obtain a quality product. Cassava starch is produced from unfermented cassava paste. The extraction of the starch is carried out in several stages:

Peeling: Peeling cassava means removing the cortex (the outer skin) and leaving only the central cylinder.

J **Washing**: The purpose of washing the peeled tubers in a basin of drinking water is to make the rest of the operation clean. Indeed, if the peeled tubers are not washed, the rest of the operations will be carried out in the dirt, which will result in a poor quality product.

J **Peeling** : The peeled and washed cassava tubers are then peeled with a device called a scraper.

Soaking and sieving: The purpose of this operation is to separate the non-rapey pieces of cassava and the large fibrous parts that form the structure from the other components of the pulp. The fractions are separated according to their size by washing the pulp with water using the sieves (the large fractions remain in the sieve while the small fractions pass to the other side). The sieving is done in two stages: a coarse sieving (with a coarse mesh sieve) to remove most of the pulp, and a fine sieving (a layer of baby cloth not yet used as a sieve) to separate the fine pulp from the starch.

J **Decanting or sedimentation**: Once the coarse particles have been separated from the fine particles, the fine particles are poured back into a large basin filled with water, while the coarse particles are wrung out and then packed in a bag for the fermentation process. After three days they are removed and dried in the sun, and used as cassava couscous for food or feed. The large basin containing the fine particles and water is left to rest for some time (this time varies according to the amount of product in the basin, the size of the basin and the amount of water, in this case it is estimated to be 5 hours). However, the aim is to leave the product to rest until the surface water becomes clear, and we are able to separate this water from the milk deposited at the bottom of the basin. We carefully remove the water that has risen to the surface. The milk remaining at the bottom of the basin still contains some water. This milk is then wrung out to extract the water.

J **Sun drying:** The paste obtained after dewatering is dried in the sun, away from birds and chickens, away from dust.

J **Sifting or blutage :** Once dried, the raw starch obtained contains granular lumps of starch and sometimes impurities. In order to obtain a fine product of good quality, we have sifted the dried product with a fine mesh sieve. The resulting product is therefore starch ready for use.

J **Packaging:** the ready-to-use starch is packaged in 0.5 litre and 1.5 litre plastic bottles. Figure 10 and 11 illustrate the process used.

Fresh tubersPeelingPeeling

Rasped   tubersDecantationDry starch

Figure 9: Cassava starch extraction process

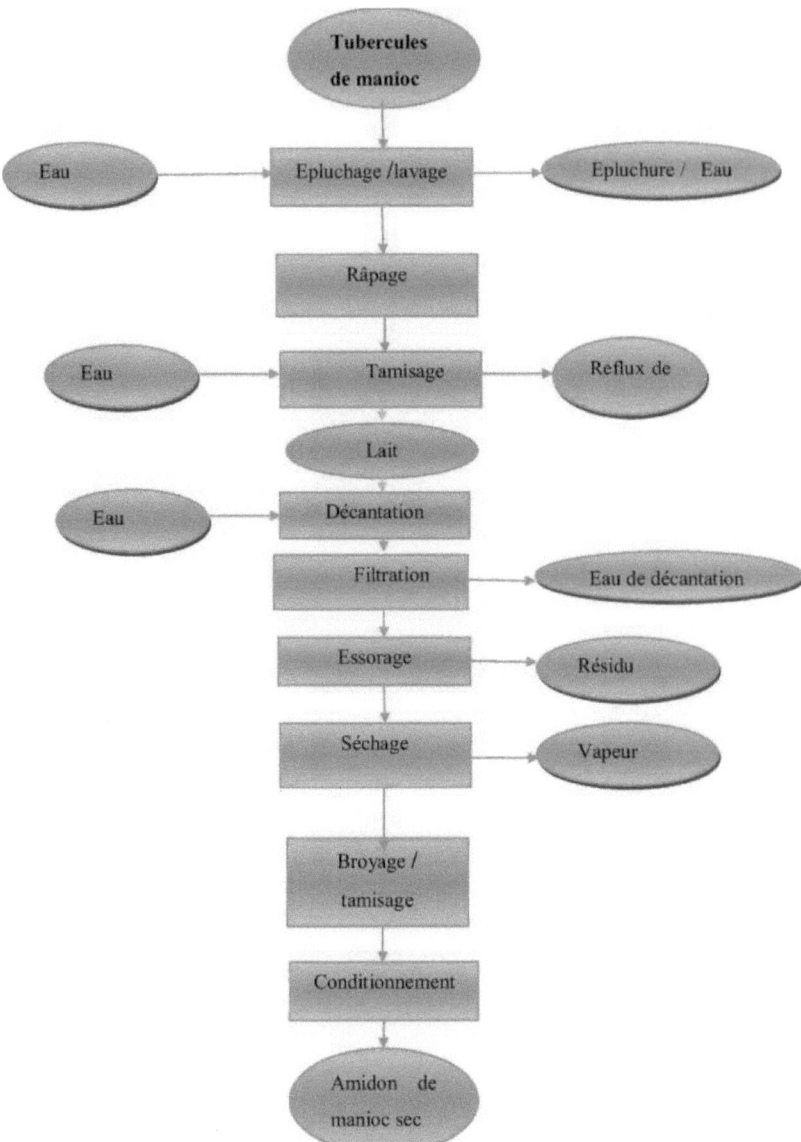

Figure 10: Synoptic diagram of the cassava starch extraction process

II.2.3.2. Extraction of chitin and chitosan

Chitin and chitosan are extracted from the shells of arthropods, in this work the extraction was done from pink shrimps or prawns which are the most common *(Palaemon fabricius)*. The extraction of chitosan is carried out according to several steps:

S Preliminary operations
- Washing : it is a question of putting the shells in a basin containing water in order to get rid of certain impurities, delimit the odors and maggots;
- Drying: exposing the shells to air for 8 hours a day to dehydrate in order to facilitate grinding and extraction;
- Grinding and sieving: pass the powder through a grinder and sieve to obtain a fine powder to facilitate extraction;

S Chitin extraction
- Demineralization: the mixture is stirred at room temperature for 8 hours using 1M hydrochloric acid at a ratio of 1:10 (W/V), the aim is to extract the mineral salts.
- Filtration and washing: distilled water is used to interrupt the reaction and remove traces of HCl;
- Deproteination: The mixture is $^{heated\ to}$ 65-90oC for 2h with 1.25M sodium hydroxide of 1:10 (W/V) solute/solvent ratio in order to extract the proteins and obtain the coloured chitin;
- Bleaching: Hydrogen peroxide is added at a ratio of 1:1 (W/V) solute to solvent, the mixture is left to stand for 2 hours to remove any residual pigments;
- Ringing : acetone is added with a ratio of solute/solvent 1/1 (w/v), the mixture is rinsed twice to see a pure white;
- Drying : the bleached chitin is dried for 24 hours at room temperature to allow evaporation of the solvents used,

S Chitosan extraction
- Deacetylation: Sodium hydroxide 50% is added at a solute/solvent ratio of 1:50 (w/v) and heated to >100oC for at least 30 min with stirring;
- Washing: washing is done with distilled water to remove traces of caustic soda;

Drying : the aim is to evaporate the traces of water and is done during 24 hours at room temperature, we obtain the dry chitosan.

The following figure 12 presents the different steps of the process used: inspired by (Tolaimate et *al* 2003; Al Sagheer et *al.* 2009)

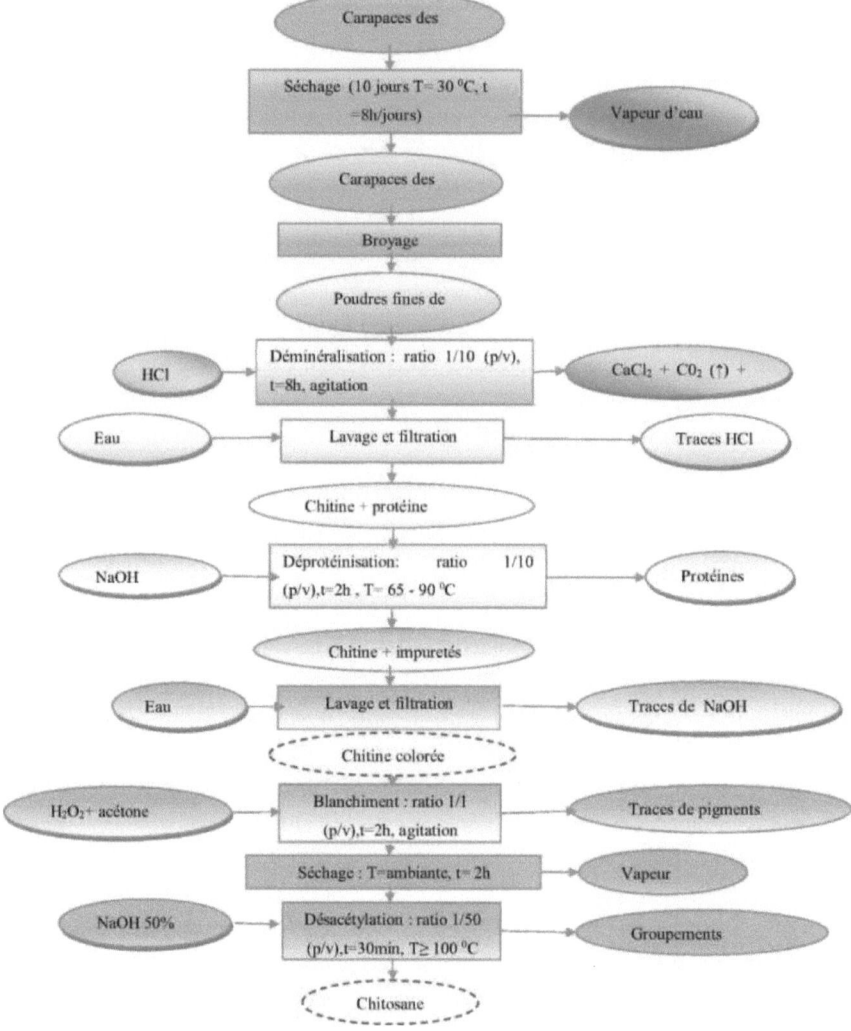

II. 2.4. Characterization Techniques

Water content

To estimate the water content of a product, 1 to 3 g of sample is taken and weighed into a jar of known weight. The jar is placed in an oven at 105 °C for 24 hours, then weighed after 30 minutes of cooling. The water content of the samples (chitin, chitosan, starch) was determined by drying the samples at 105°C for 24 h.

The water content was measured in accordance with the French standard NF ISO 11465:

The ash content is determined thermally by placing 3 g of (chitin or chitosan) in an oven at 500°C for 1 hour, cooling and weighing. The weight of the ash is expressed as a percentage of the initial weight. The ash content is evaluated according to the following formula:

$$\%ash = \frac{\text{weight of waste}}{\text{dry sample weight (g)}} - 7N \times 10°.$$

> **Purity test**

Chitosan is a large molecule (1000 to 2000 kDa,) which forms viscous solutions (Kumaret al, 2004). Chitosan exists as a water-insoluble powder. The literature review showed that aqueous solution of acetic acid is the most suitable solvent to solubilize chitosan. Unlike chitin which is insoluble in aqueous solvents. The solubility of chitosan takes place in dilute acid medium by protonation of the amine groups of chitosan in the $NH3+$ form. The test is carried out by preparing solutions of chitin and chitosan in acetic acid of 1% concentration.

II.3. FORMULATION OF BIODEGRADABLE PLASTIC FILM

To obtain the biodegradable plastic film, we proceeded with a succession of steps (figure 13) which defines the order of evolution of the product from the initial to the final stage. Thus, the process will have to go through the following stages:

> Obtaining chitosan: by the extraction method (figure 12);
> Obtaining cassava starch: by the extraction method (figure 11);
> Dissolve 5g of chitosan in 150ml of 1% acetic acid for 2 hours then sieve;
> Dissolve 15g of cassava starch and 2.5g of maize starch in distilled water then sieve;
> Add 7ml of glycerol
> Place on hot plate for 20 minutes; temperature between 70oC and 80oC; always stir the mixture over the fire
> Drips onto the glass plate while still hot
> Place in a ^{30oC} oven for 12 hours, then at room temperature for 24 hours, then demold.
> Obtaining the plastic film chitosan / cassava starch of dimension 30 / 5 in centimeters (cm)

The following figure 13 describes the process of obtaining the film by wet process. The calculation of the concentrations and quantities of the reagents are shown in annex 4.

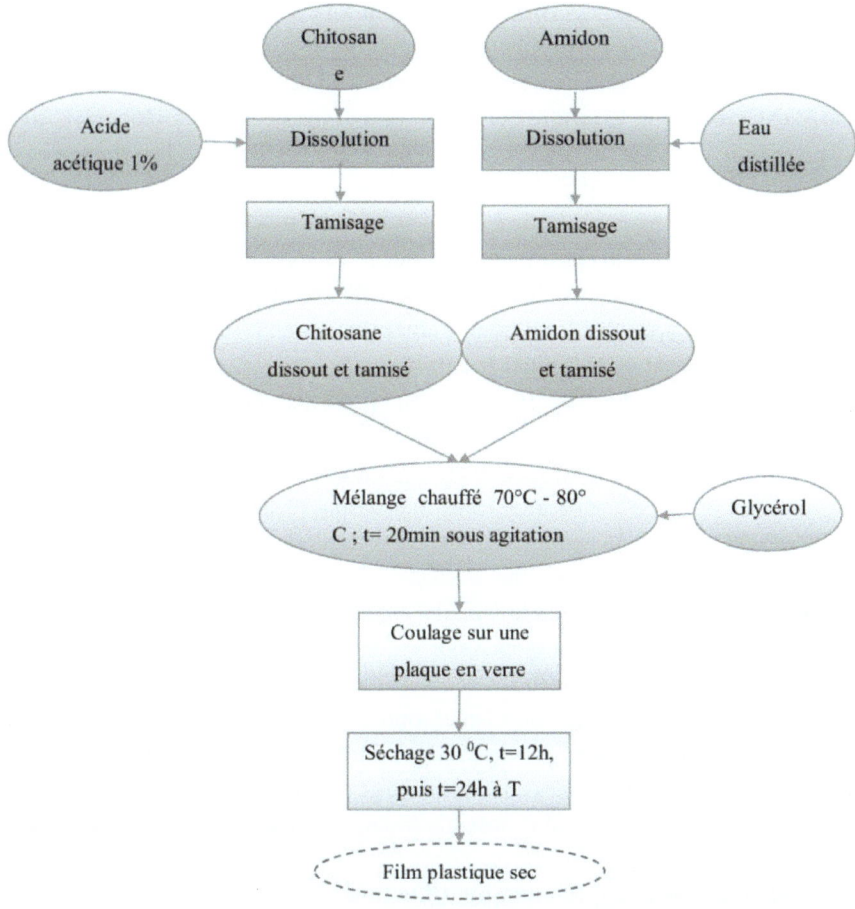

Figure 12 : schéma synoptique du procédé expérimental de fabrication du film plastique

II.4. ASSESSMENT OF FEASIBILITY

11.4.1. Cameroonian government requirements

For the purpose of this feasibility assessment, the analysis of the market from the consumer's point of view was done through a raid at the service of MINEPDED and MINCOMERCE to ensure the application of the decree of October 24, 2012. It emerges that the ban on "plastic packaging with a thickness of less than or equal to 60 microns" is still in place.

11.4.2. Study of the competition

As far as competition is concerned, the raids in the premises of MINEPDED and MINCOMMERCE give us the information that to date, biodegradable plastic packaging is not yet manufactured in Cameroon.

11.4.3. Assessment of technical feasibility

Two types of raw materials are used in the production of bioplastics: chemical products (imported) and natural products. The quantities of natural resources offered for sale depend on certain factors (rainfall, level of demand, and level of organization of the sector) which leads to a setting of collection parameters. The opinions gathered from wholesale and retail traders during the market survey made it possible to analyse the marketing circuit for cassava, shrimps and chemical products distributors in Douala.

> **Supply of bitter variety cassava**

For the supply of cassava with the roots of the "bitter" varieties, it will be done among farmers throughout the Cameroonian territory because this variety of cassava is not intended for food to avoid the conflict of food security. It would be important to look for suppliers among the large farmers for the continuous supply of this variety of cassava and if possible in the long run to create a field of bitter variety cassava.

> **Shrimp shell supply and quantification**

We will plan the supply of our shrimp shells by a collection team in the following locations:

- ❖ The naval base of the city of Douala
- ❖ The different markets of the city of Douala (Sandaga, Youpwe, Deido)
- ❖ In restaurants (le meridien, la mandarine, les mangroves...) in the city of Douala.

According to our survey on the quantity of shrimp shells produced per day in the different sites and restaurants of the place, it appears that in one day there can be more than 1000kg of shells per day in the city of Douala.

II.4.4. Testing of a production process on an industrial scale

11.4.4.1. Sizing of the bioplastic production unit

In order to determine the total production capacity of the unit, the following basic data were taken into account:

- The availability of raw material in terms of production ;
- Market demand;
- Storage constraints related to the space allocated to the storage of raw materials and finished products. Thus, the annual production capacity must be greater than the annual demand, so that the raw materials and finished products can be stored.

11.4.4.2. Choice of equipment

The choice of equipment was made in several stages:

- The definition of the need, by the list of the necessary equipments;
- Sourcing ;
- Selection of equipment.

At this level, an evaluation of all the equipment has been made with their technical characteristics (capacity, estimated lifetime, production cost, equipment price, etc.). Within the framework of a project, this step consists in analysing and selecting the specifications of the equipment suitable for the process. In selecting the equipment, the costs, the constraints related to the acquisition of the equipment, the manufacturing process and the degree of mechanization of the unit are taken into account.

11.4.4.3. Company design

The Archicad software was used to dimension the new company. The result is a 2D plan.

11.4.5. Administrative study of the project

The administrative study was done through documentary research on :

- **The administrative procedure** for the creation of a production unit in Cameroon: information on the procedure for the creation of a production unit was sought from the services of the business creation formalities centre (CFCE) in Cameroon;
- **Taxes or charges imposed** on the industrial production sector : information on taxes was obtained from the tax authorities;
- **An organization chart has** been made to represent the different departments in the company.

11.5. ECONOMIC AND FINANCIAL FEASIBILITY OF THE PROJECT

The financial evaluation is the phase of the study of a project which makes it possible to analyse whether the project is viable and under what conditions, taking into account the standards and constraints imposed on it, and on the basis of the technical and commercial studies already carried out. It consists in valuing the flows resulting from the previous studies in order to determine the profitability and financing of the project.

II.5.1. Evaluation of the investment costs

When analysing a project for the construction of a new enterprise, the first and most important decision concerns the different possibilities of investment and, if necessary, whether or not to invest. The market study helps to determine the probable volume of the product which can be put on the market and this information makes it possible to fix the minimum capacity of the factory. But in this case it is a necessity of bioplastic enterprise. For this purpose, it is a question of making a small scale estimation to cover first the city of Douala and hope to extend the project in other cities of Cameroon in the days to come.

In turn, the capacity of the plant is directly related to the investment and is reflected in the production costs. The total financial amount needed to implement a project is called "Capital Investment Cost". This investment can be made with equity capital, credits from national or international financial institutions, and suppliers. The capital needed to implement the project consists of two parts:

Fixed Capital (FC): represents the capital needed for the complete construction of the production plant with its auxiliary services and its preparation until the moment of the activity's demurrage. They represent practically the total value of the company's assets. Fixed assets may be tangible or intangible. Tangible assets include equipment (including assembly costs), buildings, ancillary facilities, etc., and intangible assets include patents, technical know-how, administrative expenses.

J Production cost (PC) : includes the capital required to achieve the level of production predicted by the technical and economic studies, after the facilities have been built. The amount of this cost varies widely, depending on the market for the products, the characteristics of the production process and the conditions regarding the origin and availability of raw materials. The fixed capital is divided into different components as shown in Table 6 below.

Table 5: Breakdown of investment expenses

Direct costs	Indirect costs
Preliminary study and analysis costs	Engineering and management
Main equipment	Construction costs
Assembly and installation costs	Contractors' fees
Networks and pipes	Unforeseen
Control equipment and instruments	

Electrical installations	
Buildings	
Ancillary services	
Land and facilities	
Start-up costs	

Once the equipment and tools have been chosen and the company's construction estimate has been drawn up, the cost of the project's investments must be determined. It consists in summing up the purchase prices of the different equipments, the setting up costs,

infrastructure costs, furniture costs and small equipment. The cost of investments is given in the following form :

Total investments and contingencies = Total general acquisitions + Contingencies on investments (5% of acquisitions).

II.5.2. Evaluation of production costs

Production costs (also known as operating costs) are the costs required to keep a given plant, production line or piece of equipment in production. Production costs are closely associated with the technical sector, which usually includes a large proportion of variable costs such as raw materials, labour and packaging. These costs are grouped in Table 7 below.

Table 6: Ranking of Production Costs (PC)

VARIABLE COSTS (CV)	FIXED COSTS (CF)
Raw material	Amortization
Direct labour	Taxes and duties
Maintenance	Interest
Supplies	Insurance
Packaging	Other costs

Thus, the production cost **(PC) = CV + CF.** The annual production cost will be multiplied by 12, taking into account the fact that a year is made up of 12 months.

II.5.3. Determination of the cost price and the selling price of the product

J **Costing**

The cost price of the good is equal to the ratio of production costs to total output. Let then :

$$PR = \frac{CP}{QP}$$

With **QP** the total production and **CP** the production cost

S **Pricing**

The selling price takes into account the investment made, the cost price and the trends of customer prices on the market.

II.5.4. Drawing up the provisional operating account

The annual operating accounts make it possible to determine the projected cash flow of the project during its life or evaluation period. To do this, it is necessary to know for each year of the project :

- The quantities produced and the unit selling price(s) ;
- Quantities of raw materials purchased and consumed and their purchase costs (ex works);
- External expenses ;
- Taxes ;
- Personnel costs ;
- Depreciation and amortization expenses; -
- Corporate tax rate.

The calculation plans of these different elements are grouped in Table 12 below

Table 7 : Methodology for the elaboration of an annual operating account

Years	N	N+1
SALES (CA)	Annual Qp x Unit selling price	
Variable expenses (CV)	Monthly CV x 12	
Fixed costs (FC)	CF monthly x 12	
REVENUBRUT OPERATING EXPENSES (EBIT)	CA- (CV+CF)	
Depreciation and amortization (DA)	IT/n; or IT= Total investment	
Interest on invested capital (15%) (IC)	DAN x 0.15	
TAXABLE INCOME (RI)	RBE-(DAN +ICN)	
Taxes (36.3%) (Ip)	IR x 0.363	
NET OPERATING INCOME (NOI)	RBE- (DA +IC + Ip)	
NET CASH (TN)	RNEN	RBEN+1
TREASURY	RNEN	TNN+ RBEN+1

Source Papin (1996);

As additional information to this table, we have :
- Qp to signify the annual production, n for the evaluation period of the project fixed at 6 for this project.
- The fixed taxes at the rate of 36,3%, takes into account the tax regiementation in Cameroon; that is to say the corporate tax IS = 33% Net Profit (RI) and the additional communal Centimes CAC = 10% of IS.

11.5.5. Determination of cash flow

Cash flow is an indicator that measures the cash flow available to a company. It is determined by the following formula: **Cash-flow= net profit + net depreciation**

11.5.6. Financial and economic profitability of the project

Profitability is a general term that measures the income that can be obtained in a particular situation. It is the common factor in all production activities. However, certain parameters must be known in order to define it. Four main criteria exist in order to evaluate it:

- The Net Present Value (NPV) ;
- Profitability Index (PI);
- Internal Rate of Return (IRR);
- The Recovery Period (DR)

4- Net present value (NPV)

The NPV is the difference between the discounted cash flows at the date and the capital invested; it expresses the potential wealth that a project can generate. It is calculated by the following formula

$$VAN = -I_T + \sum_{k=1}^{n} CF_K(1+i)^{-k}$$

With :

I_T = total initial investment,

CFk: Cash-flow or sales revenue for each year. CF here is assumed to be variable over the whole period; i: discount rate or minimum rate of return required by the company, here 15%; n: the length of the period considered estimated here to be 6 years; k: natural number between 1 and n (Ngongang, 2016).

4- Profitability Index Method

While the NPV measures the absolute benefit likely to be derived from an investment project, the profitability index (PI) measures the relative benefit, i.e. the benefit induced by 1 franc of capital invested.

$$IP = \frac{\sum_{K=1}^{n}(1-i)^{-k}}{I_T}$$

4- Recovery period (DR) OF the invested capital.

It represents the time after which the cumulative amount of discounted *cash flows* equals the capital invested. The discounted *cash flows are* made at the rate of return required by the company (cost of capital).

$$/T = {}^{\wedge}CF^*(l\text{-}i)^{D^*}$$

ks 1

(Ngongang, 2016).

4- Profitability threshold

The threshold of profitability is the turnover at which the activity of the company begins to be profitable (Ngongang, 2016).

The determination of the threshold of profitability passes by a determination of the fixed charges and the variable charges. The variable costs are the direct costs of production, while the fixed costs are management costs, personnel costs, taxes and depreciation. The following approach has been used:

$$Variable\ cost\ margin = Fixed\ costs$$

CVD=CA-CVTaux $CVD = -$

$$\textbf{CA}$$

$$\textbf{CF}$$

From or $\textit{SR - --}$(Ngongang, 2016) With:

- MCV: Margin on variable costs
- CA: Turnover
- FC: Fixed Charges
- CV: Variable expenses

The break-even **point (BP) is** deducted from the profitability threshold.

$$PM = \frac{SR}{360}$$ (Ngongang, 2016).

CHAPTER III: RESULTS AND DISCUSSION

III.1. SURVEY

A survey was conducted among the sellers of shrimp and restaurants of the place, it appears that several types of shrimp are sold namely:

- Gambas or giant prawns which are very large prawns, measuring up to 20cm in length;
- The pink shrimp or prawns are the most common, they measure 6 to 10cm;
- The red shrimp which have a common size of 10 to 17 cm in total length;
- The grey shrimp which are very small shrimps with various sizes up to 8cm.

With regard to the quantity of waste produced, the following figure 14 shows the average quantities generated per day in the various markets and restaurants in the area. For the management of these wastes it appears that those of youpwe and the naval base expose right next to the sea because of the nauseating smells that they generate to allow the wave to bring back in the water at the time of the high tide because the conservation is rather expensive in the ices if it would be necessary to keep for valorization. The others are directly evacuated in the Hysacam tanks (hygiene and sanitation of Cameroon).

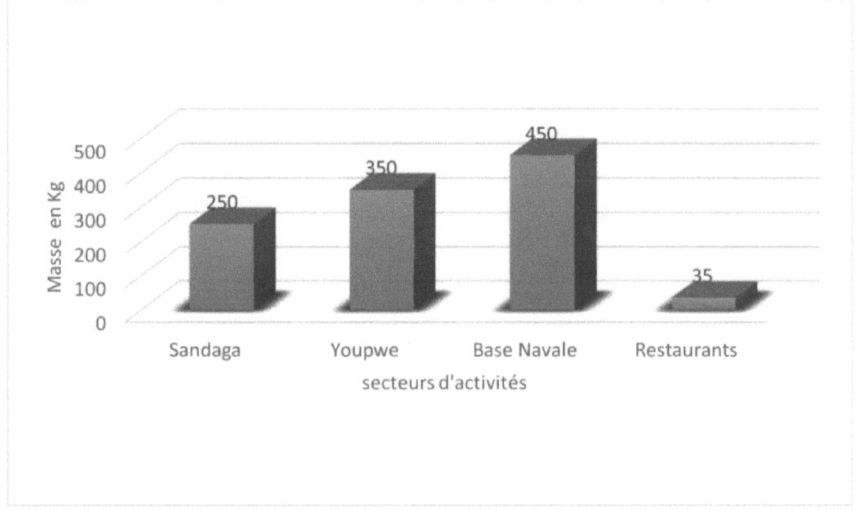

Figure 13: Waste quantities by sector of activity

The graph shows that the naval base and the Youpwe market generate more shell waste, these are the main supply areas; this is because these two areas are located right next to the river where the various fishermen are, it is from this site that the retailers, small traders and some restaurants of the city come to get supplies.

III.2. CASSAVA STARCH

Extraction from 10 kg of fresh tubers yielded a starch mass of 3.3 kg or 33%. This result is within the range predicted by the literature which is 32-35% for tubers. The starch obtained is whitish in colour as shown in Figure 16 below.

Photo 3: Starch obtained after drying

III.3. CHITIN AND PREPARATION OF CHITOSAN

III.3.1. Drying of fresh shells

After supplying the shrimp shells and pre-treating them, we proceed to drying in order to facilitate extraction by evaporating the water contained in the raw material. For 30kg of fresh shrimp shells we obtained after 10 days of drying a mass of 8kg which does not vary any more from the 8th day. The following figure 16 shows the drying curve

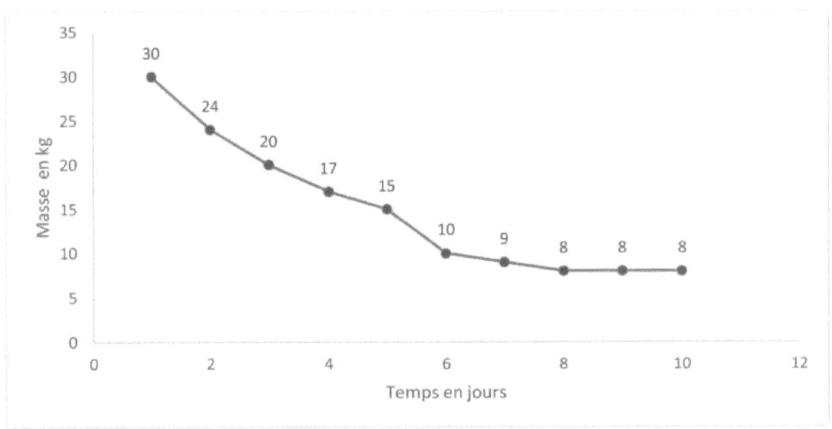

Figure 14: Drying of shells

This figure shows the quantity in kilograms of shells as a function of the number of drying days (i.e. 8 hours per day). This result could show that the adequate number of drying days is 8 days of drying in a period of moderate sunshine. After calculating the initial water content with a test sample of 25g and the mass after oven drying of 6g, we obtained 76% and compared with the final product which is 73.33%. These results are close and can be explained by the fact that the final product still has moisture compared to the oven-dried test sample. The bibliographic data predict a water content of 75% for fresh shells, but our results are around this value, which proves that the drying process was well done, despite the fact that the temperature was not controlled. We tried to follow the evolution of the mass of our sample every day until stability was reached.

Photo 4: Drying of shells

III. 3.2 Chitin and chitosan

The masses of the products obtained after shell pretreatment, demineralisation, deproteinisation, decolourisation, bleaching and deacetylation, which correspond to the mass of chitin and chitosan, are shown in Table 13

Table 8: Results of chemical extractions performed on shrimp shells

Mass of initial shell powder m0 in grams	200g
The mass of chitin extracted from 200g of shell powder.	16g
The mass of chitosan obtained after the deacetylation treatment of 16g of chitin.	9g

It can be seen from this table that the yield of chitin extraction from shrimp shells is about 8%. This yield is important, and depends on the experimental conditions (concentration of the solutions, temperature and time), it also depends on the species of crustaceans and the seasonal variations. Furthermore, the treatment of chitin for the preparation of chitosan gives a yield of 56.25%. However, according to the method given by Kurita et al. (2003) it is a process with a yield equal to 51.25%.

Following the protocol described in figure 12, for each 200g of shrimp shell powder we obtained an average of 16g of chitin. The deacetylation of this chitin gave us on average 9g of chitosan or 4.5%. It took 12 trials to obtain 100g of chitosan (i.e. 177.77g of chitin).

Literature data predict an average mass of 2.5g of chitin from 100g of shrimp shells. The following figure shows the number of extraction tests

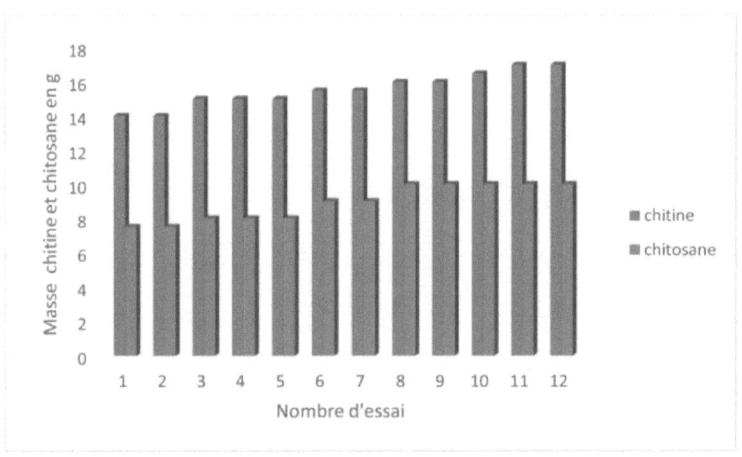

Figure 15: Extraction of chitin and chitosan

Figure 16 shows the amount of chitin and chitosan as a function of the number of trials. This difference may be due to the type of shrimp used as well as the strong interaction between proteins and chitin which could have an impact on the total protein extraction. This value may also be due to the formation of foam which influenced the temperature control and the latter is no longer fully controlled. The figures below show the appearance of the chitin and chitosan obtained.

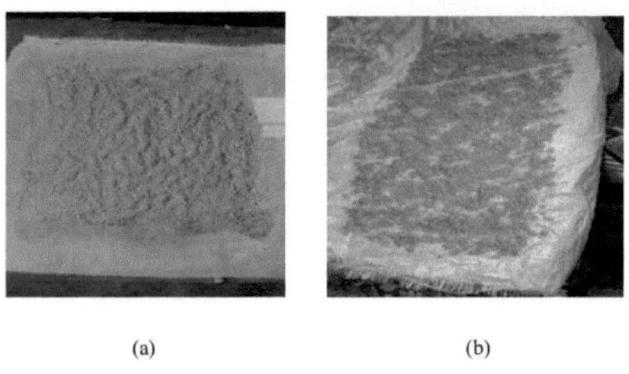

(a)   (b)

Photo 5: chitin (a) and chitosan (b)

III.3.3. Characterization of chitin and chitosan
Ash content

The products obtained after the extraction of chitin and chitosan were analyzed. To determine the ash content in a microwave oven at 500°C for 1 hour, the following table shows the results obtained:

Table 9: Chitin and chitosan ash content analysis data.

Biopolymeres	Initial mass (g)	Mass after treatment at 500°C for 1h (g)	Ash content %.
Chitin	3	0,65	**21,66**
Chitosan	3	0,5	**16,66**

The results showed that the ash levels obtained are low (21.66% for chitin and 16.66% for chitosan) which shows that the treatments on the shells to remove the mineral substances (CaCO3) are effective. Chitosan has a lower ash content than chitin and this can be attributed to the additional chemical treatments (deacetylation) that chitin underwent to obtain chitosan.

Water content

The water content of the sample was determined by drying the sample at 105°C for 24 hours until a constant mass was obtained. Table 15 shows the water content of the different samples.

Table 10: Water content of chitin and chitosan.

Biopolymeres	Initial mass (g)	Mass after oven drying at 105°C for 24 hours	Water content %.
Chitin	2	1,2	**40**
Chitosan	2	1,6	**20**

The results show that the water content of chitin is much higher than that of chitosan. This can be attributed to the amount of moisture absorbed by chitin and chitosan. The water content may vary according to the season, humidity and sunlight intensity.

> Purity test

We have observed that chitin does not dissolve in acetic acid and chitosan dissolves but not completely with some debris remaining. This result shows that the chitosan obtained is not totally pure and this can be explained by the difficulties encountered during the deproteination and deacetylation operations. According to the literature, chitin is insoluble in acetic acid and chitosan is completely soluble in acetic acid. This is because chitin in 1% acetic acid cannot be protonated due to the absence of an amine group and/or a low degree of deacetylation, whereas chitosan is complementarily soluble in 1% acetic acid and this is due to the amine group being protonated.

III.4. MANUFACTURE OF PLASTIC FILMS

III.4.1. Proportion of biopolymer

After obtaining the two biopolymers, it is necessary to formulate the plastic film. Inspired by the recent works, it is a question of finding a better proportion of two biopolymeres which gives a plastic film having a good mechanical resistance, of average thickness and not attackable by moulds and bacteria. The following table 12 presents the different proportions of the two biopolymeres as well as the results of observations:

Nº	Proportions amidon de manioc/chitosane pour une préparation de 100g	Observations visuelle	Images
1	100/0	Film très collant sur la plaque de verre non démoulable	
2	0/100	Film très cassant non démoulable	
3	50/50	Film peu élastique ayant une mauvaise résistance à l'étirement manuel très cassant et démoulable	
4	60/40	Film élastique et résistante à l'étirement manuel et peu cassant	
5	75/25	Film élastique ayant une bonne résistance à l'étirement manuel et non cassant	

Based on the different observations made, a sensory test is carried out on a sample of 10 people to reassure ourselves of the answer. The following table 13 presents the criteria and the results of the vote

Table 12: Quality IndexProportions	Critical number	Voting (10-person panel)
100/0	1= Very bad	10
0/100	2= Bad	10
50/50	6= acceptable	7
60/40	8 = good enough	8
75/25	10 = good	10

According to Table 13, the different proportions of cassava starch and chitosan that were used gave films with different behaviour. The observations represented in this table are those made on the average thicknesses (good spread) and dried in the oven and then at room temperature. All these observations allow us to conclude that when the quantity of chitosan increases, the elasticity and the resistance to stretching also increases up to a certain quantity which is the third of the cassava starch (15g cassava starch for 5g of chitosan). The same quantities (50/50) give a brittle film, which shows that a higher quantity of starch is needed to ensure the formation of the film by the wet process.

The following graph shows the appreciation for a sample of 10 people by a test of visual observation and touch. It was a question of giving a note from 1 to 10 to the different samples, we proceeded by a secret vote each participant writes his notes on 5 pieces of paper and we pass to the counting from where the result recorded on this graph.

Figure 16: Proportion of biopolymer

111.4.2. Miscibility of the mixture of the two biopolymers

After mixing the two biopolymers and heating, a single phase is observed and it is no longer possible to distinguish the original components. The chitosan mixed with the starch gives a single phase before flowing onto the glass plate. This result may be due to the fact that the two biopolymers have a very similar structure. The mixture of two biopolymers gives films with better and more elastic properties compared to the films obtained from chitosan alone and starch alone.

111.4.3. Technique for drying plastic films

Three drying techniques were used, namely oven drying at 100o C, room temperature drying and combined oven and room temperature drying. According to the methods we obtained different results.

Oven drying at 100o C for one hour

The films dried in the oven at 100o C for one hour with variable thicknesses allow us to observe the results of the following table 18:

Table 13: Drying in oven 100 ^{0C}

Film thickness	comments
Thin layers	Non-demoldable films
Average thicknesses	<ul><li>Films that are difficult to remove;</li><li>Sticky to the touch, breaking at the slightest manual stretch and showing cracks</li></ul>

The table shows that the thickness of the film influences the result. Thin films were in one layer and after drying were not demouldable due to the lack of control of thickness, temperature and drying time. After evaporation of the water, since it is a plastic, the high temperature facilitated its adhesion to the glass plate. As for the average thickness, the solution was applied in two layers. In spite of this, we still encountered difficulties during the moulding process. This allowed us to conclude that the method was not appropriate.

Drying at room temperature for 7 days

The films dried at room temperature for 7 days with variable thickness allowed us to observe the results of the following table 15:

Table 14: Drying at room temperature

Film thickness	comments
Thin layers	<ul><li>Hard to unmake movies;</li><li>Breaking at the slightest manual stretch;</li><li>Attack by bacteria and moulds.</li></ul>
Average thicknesses	<ul><li>Easy to remove films;</li><li>non-brittle and acceptable resistance to manual stretching;</li><li>No cracks, bacterial or fungal attack.</li></ul>

This drying method seems to be the best for medium thickness films but the difficulty encountered here is the time taken for drying and the attack of the plastic film by bacteria and moulds another major problem and yet chitosan according to the literature is antibacterial and antifungal. It is therefore concluded that this method is not the best one, hence the search for another alternative.

Oven drying at 30o C for 12 hours then room temperature for 24 hours
From the two thicknesses, namely fine and medium, we carried out a first drying in an oven at $^{30°C}$ for 12 hours and then at room temperature for 24 hours. This allowed us to obtain the observations shown in photo 6 and the following table

16:

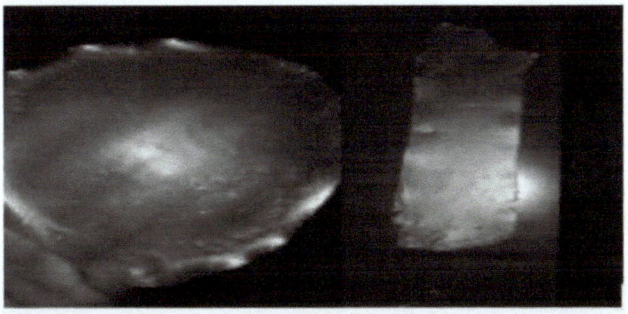

Photo 17: plastic film

Table 15: Combined drying

Film thickness	comments
Thin layers	■ Films easy to remove but breakable at the slightest manual stretching; ■ Not attacked by bacteria and moulds.
Average thicknesses	■ Easy to remove films; ■ Non-brittle ; ■ good resistance to stretching; ■ Does not have any cracks; ■ Not attacked by bacteria and moulds.

With this third method the film was easier to release and not attacked by moulds and bacteria. This may be due to the fact that bacteria and moulds attack the product as soon as it cools down after it is poured on the glass plate but once the film starts to form there is a barrier that is formed and the antibacterial activity also increases. In addition, the drying process must be moderate to avoid destroying the structure of the product. We have therefore adopted this method for the rest of this work.

III.5. TEST OF THE RESULTING BIOPLASTIC

Photo 19 shows the elongation test on biodegradable plastic. The initial length of the biodegradable plastic is 30 cm and after removal it reaches a length of 33 cm, which is the maximum strength of the biodegradable plastic that has been manufactured. The strength of a plastic strip is technically the force that it can withstand under tension per unit of cross-sectional area of the film without breaking. If you want to increase the strength of the piece of plastic, the easiest

53

thing to do is to thicken it: this would allow it to support a greater load under tension without breaking. The tensile test determines the amount of stress that each material can withstand before breaking, as well as the amount of elongation at the moment of breaking.

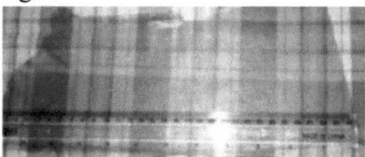

Photo 7 : élongation

III.6. FEASIBILITY STUDY
111.6.1. Commercial feasibility
111.6.1.1. Local and international competition

This is a new product on the Cameroonian territory, competing companies are based on the manufacture of plastic packaging of petrochemical nature like DEE PLAST, PLASTICAM ....In Africa GASY PLAST Malagasy company produces PVC pipes, plastic bottles and cups, but since November 2014 the company also produces biodegradable bags made of cassava husk and is based only on the local market in Madagascar, and soon they want to export throughout Africa and see their invention cross the oceans (source RFI). They produce more than 40 tons per year and hope to multiply their production to reach 360 tons. In this Malagasy company where the cassava beans are imported, a package containing 50 biodegradable plastic bags costs about 7 euro or about 4600fcfa, so a plastic bag costs 90fcfa.

III.6.1.2.   New specifications

◆**Quality of the raw material**

- **Cassava variety:** Bitter cassava, which has a high cyanide content, and is therefore not suitable for consumption. It lends itself well to production, especially since it is richer in starch.

Procurement: agricultural suppliers

- **Shrimp shell variety:** Pink shrimp or prawns are the most common, they measure 6 to 10cm and their flesh is quite fine and delicate.

Supply : Marine waste collection workers

Study of the waste collection market

◆ **Characteristic data of bioplastic**

> Thickness less than 61 microns
> Translucent, inert, easy to handle, cold resistant;
> Format A (15/10), format B (30/20)

III.6.2. good mechanical resistance.

III.6.3. Technical feasibility

III. **6.2.1 Testing of an industrial scale bioplastic manufacturing process**

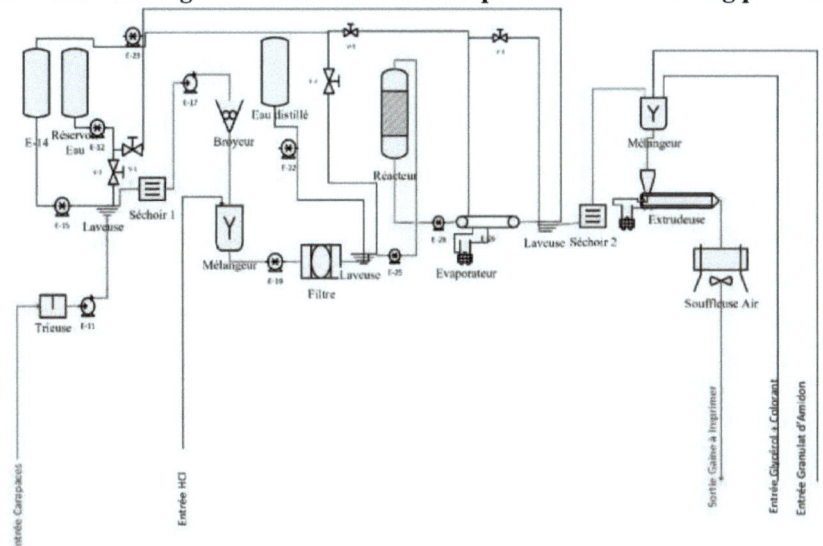

Figure 17: Diagram of the industrial process

III.6.2.2. Description of the different stages of production

This flow diagram presents the industrial manufacturing process of bioplastics with an automated biopolymer extraction system; starting with the quality control of the raw materials up to the packaging; while respecting the smallest details of quality. We will have for this manufacturing process a series of steps namely:

Preliminary operation : It is about the quality control of raw materials, cassava tubers and shrimp shells;

The shells are sorted and then crushed to facilitate the extraction process and the quality of the finished product by a gravity sorter: sorted according to size and shape, it separates the shells from the stones, small fish and other impurities then a washing with tap water is carried out then a drying at 100 ^{oC} in an oven during 1h finally with the help of a pump the dried shells are sent in a crusher to obtain fine powders.

◆ Step 1: demineralisation

In the mixing tank HCl + carapace powders are introduced and a commercial antifoam containing 10% of active silicone solution is added, followed by filtration and then washing with distilled water until neutral pH is reached, and finally the mixture is sent to the extraction reactor by means of a pump.

◆ Step 2: deproteination and deacetylation (simultaneously)

In this reactor reagents such as 1.25M NaOH for protein extraction are added. The mixture is $^{heated\ to}$ 65-90oC for 2h, NaOH 50% is added to break the acetyl groups of the chitin and heated to >100oC for at least 30min with stirring. To remove traces of residual pigments, hydrogen peroxide is added for 1 hour, followed by ringing to obtain a pure white acetone. The mixture is left to stand for 1 hour.

◆ Cassava tubers are also processed in parallel in another workshop to obtain starch using an industrial centrifuge which separates the mixture into liquid and solid in two tanks respectively.

◆ The chitosan and starch are then sent to a mixer for homogenization.

> Homogenization : it is a question of making uniform the whole of the ingredients, the mixture is introduced into a reactor of mixture then by means of a conveyor sent back in the extruder inflating;

> Extrusion inflation : During the passage of the granulates in the extruder a sheath of bioplastic is created and is formed during the passage in the various rollers;

> Printing: this is the operation that allows the visual to be printed along the entire length of the bioplastic sheath;

> Faqonnage : it is a question of giving the shape and the dimension according to the conditioning required by the company;

56

> Cutting and grouping: this operation allows to cut the ducts and to gather them in packages of the number required by the company;

> Pressing: This is the process of pressing the bundles together to give a final shape ready for sealing;

> Sealing: hermetic sealing of the different packages.

III.6.3. Choice of equipment

In order to respect the specifications and considering the technical constraints of work foreseen by the project, namely eight (08) hours of working time per day, to fix the production capacity of the factory, the proposals of offers of equipments coming from various manufacturers have oriented our choice towards a production capacity of bags per hour (5000 bags/day), to be able to satisfy the regional demand, to attack the surrounding markets and to foresee the capacity to evolve in the future

Required production and laboratory equipment

Table 16: Proposal of the equipment of the production unit for bioplastics

Cassava starch extraction	Industrial centrifuge
Film formatting	Extruder-inflator
print	Flexographic printing machine
decoupage	Automatic cutting machine
Sealing bioplastic packages	Bag sealing machine
Extraction of chitin and chitosan	Chemical reactors
Crushing of raw materials	crusher
Internal transport of materials	Conveyor trolley
Sorting shrimp shells	Gravity sorter
Supply of electrical energy in case of power failure	Generator set
Melange	Mixing tank (1000 litres)
Storage of raw materials	Storage bin
Water reserve	Water tank
Storage of non-hazardous waste	Garbage bin
Drying of samples	Oven
Microscopic observation	Microscope
Temperature control	Thermometer
Ph measurement	Ph- metre
Weighing of small quantities	Electronic scale
Weighing of large quantities	Mechanical scale
Bringing the solutions to a high temperature	Heating plate
Filling	Pails
Personal protection	PPE

The images of the selected equipment, the characteristics and the different suppliers are presented in annex 5.

III.6.4. Conception of the company

The company is located on an area of 1837 ^{m2} and provides :

❖ **A workshop for the production of plastic bags**: the operations carried out in this workshop are mainly the operations described in the process.It will rest on an area of 405 ^{m2the} construction plan is illustrative considering the needs of a standard plastics industry. It takes into account, in addition to the various workshops and work centres, the general principles of conformity of the premises as provided for by the HACCP (Hazard Analysis Critical Control Point) directives, namely:

• 4 doors minimum for the establishment:

- a door for the entry of raw materials
- a door for the entrance of the production staff
- a door for the exit of the finished products
- a door for waste disposal

  • forward motion : Because the successive working operations are planned in such a way as to ensure that the product progresses forwards, without going backwards. That is to say, from the less elaborate to the more elaborate, from the less healthy to the healthier, from the less fragile to the more fragile. In order not to disturb this order of things, the operators will be required to remain at the post to which they are assigned.

  • Non-crossing: the different production lines must not cross. They may merge (assembly of composite products, placing in a pre-cleaned packaging) or separate (processing lines for by-products obtained during the preparation of the main product).

  • separation of the hot and cold zones: The areas where the raw material is processed and the chitosan extraction tank must be clearly separated from the areas where the finished products are packaged in order to avoid thermal pollution.

  - separation of the healthy and the soiled sector: the waste produced at each stage of manufacture will be evacuated as directly as possible towards the premises devoted to their storage (dustbin).

❖ **A raw materials storage warehouse**: 50m2
❖ **General management**: 200m2.
❖ **Two changing rooms:** 30m2.
❖ **A store for finished products:** 40m2
❖ **A storekeeper's office:** 12m2
❖ **An office for the caretaker:** 12m2.
❖ **An infirmary:** 40m2.
❖ **Two toilets:** 30m2
❖ **Meeting point:** 30m2

◆ **A car park :** 100m2

III.7. Administrative study

The type of enterprise chosen is a company, namely a limited liability company (S.A.R.L.).

III.7.1. Administrative procedures

The results of the documentary research on the standard procedure for the creation of a company which was obtained from the Centre for the Formalities of Business Creation (CFCE) in Cameroon are as **follows**

- Acquisition of the authorization of the Delegate of the Urban Community of Douala ;
- Establishment of the new company ;
- Delivery by a notary of a certificate of incorporation in progress ;
- Opening of a bank account in the name of the company in formation against the delivery of a bank certificate of deposit of funds;
- Submission to the notary of the draft articles of association, the bank certificate of deposit of funds and the extract from the criminal records of the main managers of the company;
- The notary will draw up the declaration of subscription and payment of the capital;
- Collection of the signatures of the subscribers by the notary ;
- Registration of the constitutive acts and declaration of existence with the tax authorities;
- Registration in the trade and real estate credit register;
- Registration with the Statistical Service (to obtain a taxpayer number) ;
- Declaration of the workforce to the National Social Security Fund (CNPS);
Application for taxpayer card and taxpayer title at the Tax Office.
- Obtaining an environmental permit from MINEPDED
Details of the various payments to be made are set out in Annex 6.

III.7.2. Taxes or charges imposed.

> **The corporate tax IS** : still called corporate income tax, it is 33% of the income generated by the activity;
> **Additional communal cent 10% IS**
> **The patent** : it amounts to 141500 FCFA.

III.   7.3 Organization chart of the new company

Figure 20 below shows the organizational chart of the new company

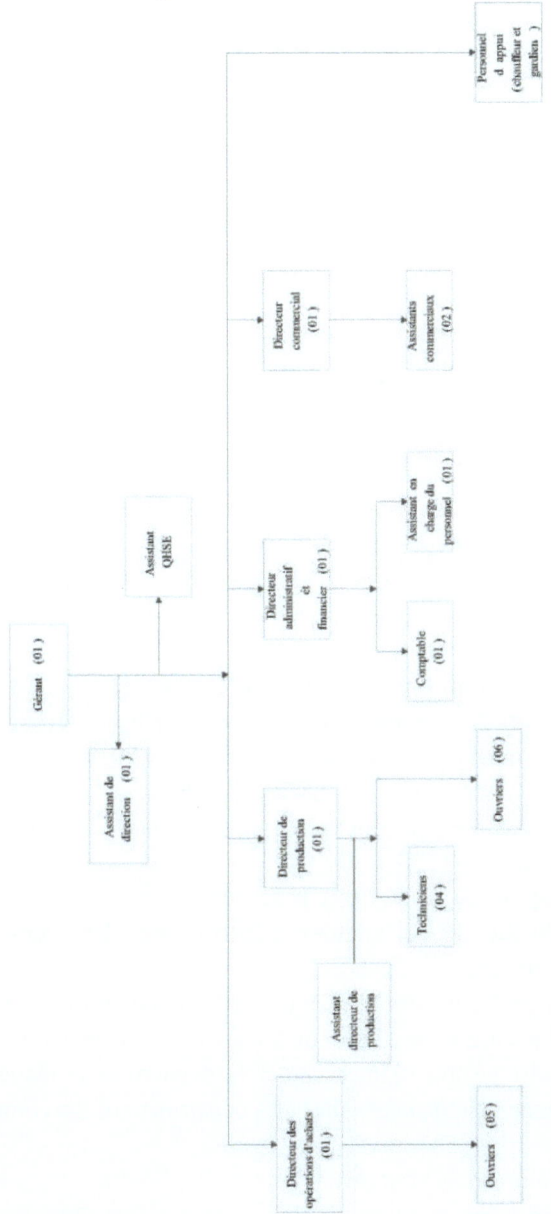

Figure 19: Organization chart

Roles of the occupants of each position

◆ **The manager** : not being a businessman, he is the legal representative of the company ; he commits the company by the acts he signs. He is responsible for the day-to-day management of the company. He defines the company's industrial policy.

◆ **Executive assistant** : it is a key position in the company. He/she manages the schedule of his/her boss and sorts the information at his/her disposal by order of importance. He plans the business meetings by choosing the most suitable places and reorganizes his agenda. He is also the representative of his superior in the outside world.

◆ **Purchasing Operations Manager** : As a manager, he is in charge of supervising the different teams that are under his responsibility. In addition, he must ensure the availability of raw material by ensuring that the supply is respected.

◆ **Production Manager** : He applies the industrial policy defined by the management. He organizes, plans and follows the production to reach the defined objectives. He/she contributes to the improvement of production processes in order to optimize productivity and guarantee customer satisfaction.

◆ **Assistant to the production manager** : He assists the latter in his various activities and replaces him in case of absence.

◆ **Administrative and financial director**: this person plays a crucial and strategic role in the company. He or she must have advanced legal skills, control of the most sensitive financial aspects and mastery of social law. He must :

- To ensure that the company's accounting system is in good shape and has sufficient cash;

- Verify the joint entries with the accountants;

- To control the tax declarations and to remain the interlocutor with the direction of the taxes;

- To have a perfect vision of human needs, to supervise the management of payrolls and to ensure the control of management and in particular of the wage bill, he will also be able to intervene in the event of crisis management.

◆ **Sales Manager** : He manages the sales department of the company, the main tasks are

- Develop the commercial strategy;

    - Implementing a policy in the company with the aim of increasing turnover and profits;

    - He participates in the design and promotion of products.

- It guarantees the network of products;
- Monitor the company's competition and keep informed of the evolution of the sector's news.

❖ **QHSE Assistant**: he/she must
- To establish and improve the documentation defining the organizational modalities (procedures, operating modes);
- Contribute to the management of risks related to the use of chemical products;
- Participate in the management of accidents and emergencies;
- To apply and educate employees on customer satisfaction, respect for the environment and the prevention of health and safety risks.

❖ **Technicians**: They have the material means that give them the ability to carry out the tasks. In this case we have a maintenance technician, bioplastic production technicians.

❖ **The accountant**: he collects, analyses and controls the expenses, income or investments of the company; he retranscribes these flows in the form of figures in the different accounts of the accounting plan.

❖ **Assistant in charge of personnel:** he/she is in charge of recruitment, follow-up of the training plan; administrative management of employees.

❖ **Commercial assistant :** real intermediary between the customer and the company, he is in charge of processing the orders, elaborating the estimate, he ensures the follow-up of the customers and their file.

❖ **The workers :** They are in charge of the execution of the various tasks of production and supply under the coordination of their various responsible.

❖ **Driver :** he is in charge of transporting the company's raw materials and finished products.

❖ **The caretaker :** he ensures the security of goods and people, the cleaning and maintenance of the common areas, the reception, the information and the orientation of the visitors.

III. 8. financial studies of the project

The financial evaluation is the phase of the study of a project which allows to analyze if this project is viable, and in which conditions, taking into account the standards and the constraints which are imposed to him. For this purpose, it is necessary to determine the *cash-flow,* calculate the Net Present Value, the payback period, the profitability threshold in order to conclude on the viability of the project.

III.8.1.   Initial investment costs and working capital for the first year

All investments are made before the start-up of the unit: all investments are made at the beginning of the first year of activity (year 1)

Initial investment costs in equipment

Table 17: Initial investment cost of equipment

Elements	Quantity	Unit price(in FCFA)	Amount (in FCFA)
Establishment costs			
Incorporation fees			508 777
Assembly: technician's fees, transport costs			6 500 000
Other pre-operational costs			1 000 000
Total set-up costs			**8 008 777**
Infrastructure			
Land			30 000 000
Plant construction and drilling			54 965 940
Total infrastructure			**84 965 940**
Production and laboratory equipment			
Industrial centrifuge	1	12 025 928	12 025 928
Automatic printer - cutter	1	8 133 904	8 133 904
Bag sealing machine	1	200 000	200 000
Chemical reactor mixer	1	6 500 000	6 500 000
Shredder	1	825 000	825 000
Conveyor trolley	1	60 000	60 000
Gravity sorter	1	250 000	250 000
Generator set	1	650 000	650 000
Mixing tank	1	165 000	165 000

Storage bin	1	85 000	85 000
Water tank	1	100 000	100 000
Garbage bin	1	140 000	140 000
Oven	1	4 198 144	4 198 144
Microscope	2	148 129,50	296 259
Thermometer	2	50 000	100 000
Ph- metre	2	100 750	201 500
Electronic scale	2	18 307,70	36 615
Mechanical scale	2	120 000,00	240 000
Glassware	1	250000	250 000
PPE	1	1 000 000	1 000 000
S/total CAF value			**35 457 350**
Customs duty (10% CIF value)			3 545 735
VAT/ imported products (19.25% of the turnover of imported material)			1 365 108
Total production equipment			**40 368 193**
Transport equipment			
Distribution vehicle and supply	1	7871520	7 871 520
Vehicle (tricycle)	2	1200000	2 400 000
Total transport equipment			**10 271 520**
Office furniture and equipment			
Binder	5	30 000	150 000
Office	8	92 000	736 000
Chair	20	15 000	300 000
Cabinet	2	40 000	80 000
Total office furniture and equipment			**1 266 000**
Computer equipment			
Computers	5	150 000	750 000
Printers	5	75 000	375 000
Total computer equipment			**1 125 000**
Small equipment			
Pails	15	2 000	36 000

Covers	5	5 000	7 200
Bins	10	5 000	40 000
Boxes	5	7 000	6 000
Total small equipment			**89 200**
Total			**146 094 630**
Unforeseen=5% of acquisitions			7 304 732
Grand total			**153 399 362**

This list was elaborated from the prices provided by the supplier companies on line and the estimate made by engineer KANA memoire end of studies CIGE (2014) with the civil engineering companies, ENER-BTP, (for the estimate of the construction of the factory), drilling (by a French company of the sale of the industrial material, MATCO France) and by the local stores of distribution. The total amount of real estate amounts to **146 094 630F CFA**, with a 5% margin for unforeseen expenses, i.e. a total of **153 399 362F CFA.**

Initial investment costs in raw materials necessary for the start-up of the activities

Table 18: Raw material and reagents for the first year

Designation	Quantity	PU	Total price (in fcfa)
Hydrochloric acid 2000 L Hydroxide		3000	6000000
sodium	5500Kg	2000	11000000
acetone peroxide	2500 L	2000	5000000
of hydrogen	2500 L	1500	3750000
dyes	76,67 L	5000	383350
shells	360000Kg	5	1800000
Cassava	250000Kg	20	5000000
TOTAL			32933350
Unforeseen			1646665

TOTAL PM and Reagents   34580015

The above table shows the raw materials and reagents required for the operation of the enterprise during the first year of activity. The total cost of the raw materials quantified is **CFAF 32 933350**, with contingencies estimated at 5% of the total cost of the raw materials, i.e. **CFAF 1646665**, giving a total annual cost of CFAF **34580015**. Taking into account the storage capacity of the enterprise, the availability of funds and the shelf life of the products, the enterprise could not buy all its annual raw material in a single instalment, but it will make monthly purchases of its raw material; it will pay an average of **2881167.917 F CFA** per month for the purchase of its raw material. In sum, Tables 22 and 23 show that the initial investment in the project, representing the real estate, raw materials and reagents needed to run the enterprise for one month, amounts to CFAF 156280529.**9**.

The following table 20 shows the different depreciation of the selected equipment.

Table 19: Calculation of the different depreciation

Headings	Value	Duration of life (year)	Year 1	Year 2	Year 3	Year 4	Year 5	Year 6	Residual value
Set-up costs	8 008 777	5	1 601 755	1 601 755	1 601 755	1 601 755	1 601 755		0
Buildings, workshops and offices	54 965 940	20	2 748 297	2 748 297	2 748 297	2 748 297	2 748 297	2 748 297	38 476 158
Computer equipment	1 125 000	4	281 250	281 250	281 250	281 250	0	0	0
Small equipment	89 200	1	89 200	0	0	0	0	0	0
Equipment of production	40 368 193	6	6 728 032	6 728 032	6 728 032	6 728 032	6 728 032	6 728 032	0
Office furniture and equipment	1 266 000	4	316 500	316 500	316 500	316 500	0	0	0
Transport equipment	10 271 520	5	2 054 304	2 054 304	2 054 304	2 054 304	2 054 304		0
Unforeseen 5	7 304 732	6	1 217 455	1 217 455	1 217 455	1 217 455	1 217 455	1 217 455	0
TOTAL	123 399 362		15 036 794	14 947 594	14 947 594	14 947 594	14 349 844	10 693 784	38 476 158

		108 362 568	108 451 768	108 451 768	108 451 768	109 049 518	112 705 577
Residual value	123 399 362						

To evaluate the cost of production, the quantity of electricity consumed and the cost of raw materials were taken into account. The following table shows the annual price:

Table 20: Evaluation of the production cost

Evaluation of production and distribution costs

Libelle	Price per unit (FCFA)	Unit	Annual quantity	Year 1	Year 2	Year 3	Year 4	Year 5	Year 6
Purchasing	19		1 820 000	34 580 000	34 580 000	34 580 000	34 580 000	34 580 000	34 580 000
Electricity	15	110KWh		1 650 000	1 650 000	1 650 000	1 650 000	1 650 000	1 650 000
Total				36 230 000	36 230 000	36 230 000	36 230 000	36 230 000	36 230 000

III.8.2. Unit costing and pricing of products
III.8.2.1 Production rate

> Let's fix the quantity of type A plastic bags

Quantity of bags produced per hour If 1min -> 7 bags of type A, 1h -> *420 bags*

Quantity of bags made per day, per month and per year: 8 h -> 3360 bags / day, 1 month -> 100800 bags / month, 1 year -> 10080 x 12 = 1209600 bags/year.

> Let's set the amount of plastic bags of type B

Quantity of bags produced per hour If 1min -> 3 type B bags, 1h -> 180 bags

Quantity of bags made per day, per month and per year: 8 h -> 1440 bags / day, 1 month -> 43200 bags / month, 1 year -> 4320 Ox 12 = 518400 bags/year

> Let's set the quantity of unused plastic packages

In order to complete our production schedule at 1820000 bags/year. We have 1820000 - 1728000 = 92000sacs/year.

III.8.2.2. Unit cost of plastic bags

The cost of a product depends on the variable and fixed costs. The costs are recorded in table 22

Table 21: Unit cost

Cost of ownership	Year 1	Year 2	Year 3	Year 4	Year 5	Year 6
Production and distribution costs (FCFA)	36 230 000	36 230 000	36 230 000	36 230 000	36 230 000	36 230 000
Staff costs (FCFA)	32 580 000	33 231 600	33 883 200	34 534 800	35 186 400	35 838 000
General management costs (FCFA)	325 500	570 500	570 500	570 500	570 500	570 500
Total	**69 135 500**	**70 032 100**	**70 683 700**	**71 335 300**	**71 986 900**	**72 638 500**
bags	1820000	1911000	2002000	2093000	2184000	2275000
Units produced A+B (packages)	1 820 000	1 911 000	2 002 000	2 093 000	2 184 000	2 275 000
Unit cost (FCFA)	**38**	**37**	**35**	**34**	**33**	**32**

70

III.8.2.3 Pricing of plastic packages

Given that there are two formats Format A (15/10), Format B (30/20) for production, the production capacity is set at 5000 bags per day, depending on the average unit cost the selling price of different formats is set. The following table 23 shows the different prices per year and the estimated turnover.

Table 22: Sales Projections

Elements	Year 1	Year 2	Year 3	Year 4	Year 5	Year 6
Units produced A+B (bags)	1 820 000	1 911 000	2 002 000	2 093 000	2 184 000	2 275 000
Average unit cost	38	37	35	34	33	32
Cost price A+B	69 135 500	70 032 100	70 683 700	71 335 300	71 986 900	72 638 500
Units sold A (bags)	1 274 000	1 337 700	1 401 400	1 465 100	1 528 800	1 592 500
Unit selling price A	45	45	45	50	50	50
s/Turnover A	57 330 000	60 196 500	70 070 000	73 255 000	76 440 000	79 625 000
Units sold B (bags)	546 000	573 300	600 600	627 900	655 200	682 500
Unit selling price B	55	55	60	60	60	60
s/Turnover B	30 030 000	31 531 500	36 036 000	37 674 000	39 312 000	40 950 000

71

III.8.2.4. Determining cash flow

The turnover represents what the company sold per year. The beginning of any activity is not obvious because the sales are not yet huge. It is therefore preferable to make an increasing estimate of the turnover. The following table 24 shows the determination of the cash flows over six years.

Table 23: Determination of cash flow

Elements	Year 1	Year 2	Year 3	Year 4	Year 5	Year 6
Turnover	87 360 000	91 728 000	106 106 000	110 929 000	115 752 000	120 575 000
PM and Reagents	36 230 000	36 230 000	36 230 000	36 230 000	36 230 000	36 230 000
Miscellaneous management fees	325 500	570 500	570 500	570 500	570 500	570 500
Personnel costs	32 580 000	33 231 600	33 883 200	34 534 800	35 186 400	35 838 000
Gross operating surplus (GOI)	18 224 500	94 155 900	107 882300	112 053700	116 225100	120 396500
Depreciation and amortization	15 036 794	14 947 594	14 947 594	14 947 594	14 349 844	10 693 784
Operating result	**3 187 706**	**79 208 306**	**92 934 706**	**97 106 106**	**101 875 256**	**109 702 716**
Corporate tax 33%.	1 051 943	26 138 741	30 668 453	32 045 015	33 618 835	36 201 896
Net result	**2 135 763**	**53 069 565**	**62 266 253**	**65 061 091**	**68 256 422**	**73 500 819**
Cash flow	17 172 557	68 017 159	77 213 847	80 008 685	82 606 265	84 194 604

III.8.2.5.Assessment of project viability

The net present value and the recovery period were determined for a discount rate of 15% (reference discount rate in Cameroon) and the data are contained in the following table:

Table 24: Project Sustainability

Years	1	2	3	4	5	6
Cash flow (CF)	17 172 557	68 017 159	77 213 847	80 008 685	82 606 265	84 194 604
(1+i)⁻ⁿ	0,86956522	0,75614367	0,65751623	0,57175325	0,49717674	0,43232760
CF(1+i)-n	14 932 658	51 430 744	50 769 358	45 745 225	41 069 913	36 399 651
^Accumulated cash	14 932 658	66 363 402	117 132 760	162 877 985	203 947 899	240 347 549
Initial investment			-153 399 362			
VAN			86 948 187			
DR			3 year(s) 11 months			
IP			1,56			

The NPV is positive which means that the project is viable. The value of the payback period shows that after 3 years and 11 months there will be a return on investment. Thus, the advantage induced by 1 franc of the invested capital is 1,56francs.

73

III.8.2.6 Determining the profitability threshold

The following table shows the values of the breakeven point and the break-even point (the threshold to be reached for an option to be profitable)

Table 25: Determination of the profitability threshold

Elements	Year 1	Year 2	Year 3	Year 4	Year 5	Year 6
Turnover	87 360 000	91 728 000	106 106 000	110 929 000	115 752 000	120 575 000
Variable expenses	69 135 500	70 032 100	70 683 700	71 335 300	71 986 900	72 638 500
Margin on variable costs	18 224 500	21 695 900	35 422 300	39 593 700	43 765 100	47 936 500
M/CV rate	21%	24%	33%	36%	38%	40%
Fixed costs	15 036 794	14 947 594	14 947 594	14 947 594	14 349 844	10 693 784
SR				259 921 604		
Neutral point				2 year(s) 5 months		

The business will start to be profitable by the middle of the third year.

III.9. SOCIAL INTEREST OF THE PROJECT

From the social point of view, the bioplastic production unit from shrimp shells has the following advantages

- The production unit will provide employment, which will reduce unemployment;
  - The use of shrimp waste for the production of bioplastic will reduce a considerable amount of waste discharged into the environment;
  - The use of bioplastic will reduce greenhouse gas emissions and protect the environment.

CONCLUSION AND OUTLOOK

This thesis work focused on the formulation of a bioplastic based on two biopolymers: cassava starch / chitosan and on the creation of a manufacturing unit on an industrial scale from the devaluation of a technical, financial and commercial feasibility. The first objective was to use materials from biomass, in this case cassava starch and chitosan, to develop new materials that are more environmentally friendly so that after use, they degrade rapidly in a natural way. In the first part of this work, starch was extracted from cassava tubers, which served as the main raw material for this study, and chitosan was extracted from shrimp shells, which served to improve the impermeability and mechanical properties of the films in order to overcome the main drawbacks of the starch materials, which are their hydrophilic character and their poor mechanical properties. In the second part of this work, a technical, financial and commercial feasibility study for the profitability of our project was carried out. The study shows that it is necessary to implement the more environmentally friendly plastic films from these two biopolymers. Thus, a trial process for the production of biodegradable plastic films on an industrial scale was proposed. The financial evaluation allowed to estimate the global investment cost of the first year of operation at **189 629 362 F CFA** and that after **3 years and 11 months** there will be a return on investment. The analysis of the project has led to a positive net present value, thus reflecting the viability of the project. The profitability threshold is **259 921 604 FCFA** and the profitability index is **1.56 FCFA**. The legal status chosen for this future enterprise is the limited liability company, a choice motivated by flexibility in its management, credibility and also sustainability.

In perspective to this work,

> Further characterization of the chitosan obtained in order to compare it with the one on the market;

> The addition of reinforcing fillers in order to make a physicochemical and mechanical characterization of the obtained film to be able to make a comparison with other biodegradable plastics according to the standard EN 13432 before implementing the project;

> Studies for bank loans must also be carried out to know who to lend to, when, for what possible amount and at what interest;

> An environmental and social impact study should be carried out on the site of the production plant in order to determine the impact of such a project on the environment.

BIBLIOGRAPHIC REFERENCES

A.O.A.C. (1980) : Official Methods of Analyses of the Association of Official Analytical Chemists,Association of Official Analytical Chemists, Washingtin : DC.

Akiyama M., Tsuge T., Doi Y. (2003). Environmental life cycle comparison of polyhydroxyalkanoates produced from renewable carbon resources by bacterial fermentation. Polym. Degrad. Stab. 80, p. 183-194.

Al Sagheer FA, Al-Sughayer MA, Muslim S, Elsabee MZ (2009). Extraction and characterization of chitinand chitosan from marine sources in Arabian Gulf. *Carbohydr Polym,* 77(2):410-419.

ANDRIEUX G., (2004). La filiere frangaise des co-produits de la peche et de l'aquaculture: état des lieux et analyse. Etudes de l'Ofimer, pp 63.

JOINT ORDER N°004 MINEPDED/MINCOMMERCE OF 24 OCTOBER 2012 Regulating the manufacture, import and marketing of non-biodegradable packaging.

Berth, G., H. Dautzenberg , et M. G. Peter. (1998). Physico-chemical characterization of chitosansvarrying in degree of acetylation. Carbohydrate Polymers, 36 ,205-2 18.

Biot, J.B. (1844). Note on the phenomena of polarization produced through the feculent globules. C. R. Hebd. SceancesAc. *Sci,* 18, 795-797.

Braconnot, H. (1811). Recherches analytiques sur la nature des champignons. Annales de Chimie, 79, 265-304.

Brugnerotto, J., J. Lizardi, F.M. Goycoolea, and M. Rinaudo. (2001). An infrared investigation in relation with chitin and chitosan characterization. Polym., 42, 3569-3580.

Buleon, A., Colonna, P. and Leloup, V. (1990). Starches and their derivatives in the cereal industries. *Industries Alimentaires et Agro-Alimentaires*, 107(6), 515-532.

Chandra, R. and Rustgi, R. (1998). Biodegradable polymers. *ProgPolymSci,* 23, 1273- 335

Chatelet, C., O.Damour, and A. Domard. (2001). Influence of the degree of acetylation on some Biological properties of chitosan Films. Biomater., 22, 261-268.

CHEFTEL J-C. & CHEFTEL H. (1976). Introduction a la bichimie et a la technologic des aliments Volume 1. *Tech & Doc Lavoisier,* Paris - France. 381p.

Chen, R.H., J. R. Chang, and J. S. Shyur. (1997). Effects of ultrasonic conditions and storage in acidic solutions on changes in molecular weight and polydispersity of treated chitosan. Carbohyd. Res., 299, 287-294

DUMAY, J.; DONNAY-MORENO, C.; BARNATHAN, G.; JAOUEN, P.; BERGE, J.P., (2006). Improvement of lipid and phospholipid recoveries from sardine (Sardinapilchardus) viscera using industrial proteases. *ProcessBiochemistry,* 41, 2327-2332.

Emmanuel K. (2014). Montage et valorisation d'un projet de création d'une entreprise cosmetique, *memoiroutenu* a l'ENSAI de Ngaoundere, 74 pages.

Fang,N., Chan,V., Mao,H.Q. &Leong, K.W. (2001). Interactions of phospholipid bilayer with chitosan: effect of molecular weight and pH.Biomacromol., 2, 1161-1168.

Heux, L., Brugnerotto, J., Desbrieres, J., Versali, M.F. &Rinaudo, M. (2000). Solid state NMR for determination of degree of acetylation of chitin and chitosan. Biomacromol., 1, 746751.

Jaouen, D. (1994). Chitin, Chitosan and Derivatives. PhD These in Pharmacy, University of Angers, Angers (France).

Kasaai, M. R., J. Arul, S. L. Chin et G. Charlet. (1999).The use of intense femto second laser pulses for the fragmentation of chitosan. J. Photochem. Photobiol., A: Chem., 120, 201-205.

Khan, T. A., Peh, K. K., and C'ing, H. S. (2000). Mechanical bioadhesive strength and biological evaluation of chitosan films for wound dressing. Journal of Pharmaceutical and Pharmaceutical Science, 3(3) 3003-3371.

KIM, S.K.; RAJAPAKSE, N.; SHAHIDI, F., (2008). Production of bioactive chitosan oligosaccharides and their potential use as nutraceuticals. In: C.S. Barrow, F (Ed.) Marine nutraceuticals and functional foods. Nutracetical Science and Technology, New York, pp. 183196.

Kurita K. (2006). Chitin and chitosan: Functional biopolymers from marine crustaceans, Marine Biotechnology. (8) 203-226.

Lang, G., Clausen, T. (1989). The use of chitosan in cosmetics. In: Skjak-Braek, G., Thorleif Anthosen, T., Standford, P. (Eds.), Chitin and Chitosan. Sources, Chemistry,Biochemistry. Physical Properties and Applications. Elsevier Applied Science,London and New York, pp. 139-147.

Legros, N., Chapleau, N. and Li, H. (2011). Plastics processing and biosource materials. Colloque quebecois sur les bioplastiques compostables, pages 200-203.

LELOUP V., COLONNA P. & BULEON A., (1991). Influence of amylose-amylopectinratio on gel properties. *J. CerealSci.* 13: 1-13.

Liu, H., Xie, F., Yu, L., Chen, L. et Li, L. (2009). Thermal processing of starch-based polymers. *Progress in Polymer Science,* 34(12), 1348-1368.

Law n°96/12 of 5 August 1996 on the framework law on environmental management

Maghami, G. G., and G. A. F. Roberts. (1988). Evaluation of the viscometric constants for chitosan. Die MakromolekulareChemie, 189 (1), 195-200.

Massive online open course Collette and Marketing of Plastic Waste Edition 2 BELLOMAR LEARNING 2019

Mekonnen, T., Mussone, P., Khalil, H., &Bressler, D . (2013). Progress in bio-based plastics and plasticizing modifications. [10.1039/C3TA12555F]. Journal of Materials Chemistry A, 1(43), 13379-13398.

Miles, M.J., Morris, V.J. and Ring, S.G. (1985). Gelation of amylose. *Carbohydrate Research,* 135(2), 257-269.

Mutungi, C., Onyango, C., Doert, T., Paasch, S., Thiele, S., Machill, S., Jaros, D. and Rohm, H. (2011). Long- and short-range structural changes of recrystallised cassava starch subjected to in vitro digestion. *Food Hydrocolloids,* 25(3), 477-485.

Muzzarelli, R.A.A. (1977). Chitin; Pergamom: Oxford.

Ngongang. (2016). *Corporate strategic.* Support de cours. E.N.S.A.I., University of Ngaoundere, 63 pages.

Nielsen L.E. (1974). Mechanical properties of polymers and composites II. Eds Marcel Dekker,New-York (USA).

No, H.K. and Meyers, S.P. (1995). Preparation and characterization of chitin and chitosan. A review. J. Aquatic Food Prod. Tech., 4, 27-52.

No, H.K., S.P. Meyers, and K.S. Lee. (1989). Isolation and characterization of chitin from crawfish shell waste. J. Agric. Food Chem., 37, 575-579.

Papin R. (1996), Strategie pour la creation d'entreprise : création, reprise, developpement, Dunod. 402 pages.

Ravi Kumar, M.N.V. (2000). A review of chitin and chitosan applications. React. Function. Polym., 46, 1-27.

Rhim, J.W. et Perry, K.W. (2007). Natural Biopolymer-Based Nanocomposite Films for Packaging Applications. *Critical Reviews in Food Science and Nutrition,* 47(4), 411 — 433

Seng, J.M. (1988). Chitin, chitosan and derivatives: new perspectives for the industry. Biofutur, 9, 40-44.

Shahidi, F. and Abuzaytoun, R. (2005). Chitin, chitosan, and co-products: chemistry, productions, applications, and health effects. Adv. Food Nutr. Res., 49, 93-135.

Shahidi, F., J.K.V. Arachi, and Y.J. Jeon. (1999). Food applications of chitin and chitosan. Trends Food Sci. Tech., 10, 37-51.

Souza, A.C., Benze, R., Ferrao, E.S., Ditchfield, C., Coelho, A.C.V. et Tadini,

C.C. (2011). Cassava starch biodegradable films: Influence of glycerol and clay nanoparticles content on tensile and barrier properties and glass transition temperature. *LWT - Food Science and Technology,* 46(1), 110 - 117.

TCHAKOUNTE Josiane (2015). Cameroon Tribune Thursday 14 May 2015, 14:43:50 reaction.

Tolaimate, A., J. Desbrieres, M. Rhazi and A. Alagui. (2003). Contribution to the preparation of chitosans with controlled physico-chemical properties. Polym., 44, 7939-7952.

Valbiom, (2011). http://www.valbiom.be/ files/gallery/amidonpla 20111297333283.pdf
(15/07/2019)

Valentin D. and Thomas V. December 13, 2010 *"Chitosan - a biopolymer of the future for antimicrobial papers" [online] accessed June 30, 2019;*

Yang,B.Y. & Montgomery, R. (2000). Degree of acetylation of heteropolysaccharides. Carbohydr. Res., 323, 156-162

Annex 1: Extract from joint order n°004 minepded/mincommerce of 24 October 2012

SECTION I
OF PLASTIC PACKAGING

Article 7 - (1) The manufacture, import, possession and marketing or distribution free of charge of non-biodegradable plastic packaging with a low density of less than or equal to 60 microns (1 micron is equal to 1/1000 mm) as well as the granules used for their manufacture are prohibited.

(2) the production, import, detention, marketing of non-biodegradable plastic packaging of more than 60 microns and of the granules used for their manufacture are subject to obtaining an environmental permit referred to in Article 4 above.

Article 8 - (1) The thickness, the formulation, the biodegradability or not, the precise name and address of the manufacturer shall be indicated on the plastic packaging manufactured or imported in accordance with the regulations in force.

(2) The particulars referred to in paragraph 1 above shall be clearly visible and easily legible to facilitate identification and classification.

Article 9 - It is strictly forbidden to burn plastics in the open air, to throw them in nature or to bury them.

CHAPTER III
MISCELLANEOUS, TRANSITIONAL AND FINAL PROVISIONS

(1) Any manufacturer, importer or distributor of non-biodegradable packaging shall have eighteen (18) months from the date of signature to comply with the provisions of this Order.

(2) After the period mentioned in paragraph 1 above, the competent authorities shall proceed to the control, seizure and destruction of non-biodegradable packaging at the expense of the promoter.

Article 13 - The administrations in charge of the environment and trade are responsible, each in its own area, for the application of this order.

Article 14 - The present decree shall be registered, published according to the emergency procedure, and then inserted in the official journal in French and English.

Yaounde, 24 October 2012

THE MINISTER OF COMMERCETHE MINISTER OF
THE ENVIRONMENT, THE
(e) Luc Magloire MBARGAPROTECTIONDELA
ATANGANANATUREAND
SUSTAINABLE DEVELOPMENT
(e) HELE Pierre

1-Fiche de préparation de l'acide chlorhydrique

Nous devons convertir l'acide chlorhydrique car notre acide à une concentration de 33% or nous voulons travailler avec une concentration de 1M.

Ici, le volume que nous voulons avoir est $V_f = 500$ ml. %P représente le pourcentage massique de pureté qui correspond à 33%.. d=densité relative en kg/l, C = concentration molaire en mol/l, M= masse molaire en g/mol

①

Or, d'après le cours de chimie on a : $C = \frac{\%P \times d \times 1000}{100 \times M}$; $X = \frac{Cf.Vf.Mi.100}{X \times d}$

$\longrightarrow C = \frac{33 \times 1,19 \times 1000}{36,5 \times 100}$; $C = 10,75$ M

Nous savons que : $C_iV_i = C_fV_f$ $\longrightarrow V_i = \frac{CfVf}{Ci}$ AN : $V_i = \frac{1 \times 500}{10,75}$;

$V_i = 46,47$ **mL** est la valeur à prélever de la solution mère et à introduire dans une fiole jaugée de 500mL puis compléter jusqu'au trait de jaugé à l'eau distillée.

2-Préparation de l'hydroxyde de sodium pour une concentration 1,25M

Trouvons la masse de cristaux d'hydroxyde sodium nécessaire pour C = 1,25 M ;

V = 500 ml = 0,5 l

Nous savons que : $C = \frac{m}{M V}$ $\longrightarrow$ **m = CVM** ②

83

M = 40g ⟶ m = 1,25 ×0,5 × 40   d'où m = 25g à peser pour une solution de 500mL ; toujours dans la fiole jaugé

3- Fiche de préparation de soude pour une concentration de 50%.

Trouvons la masse de cristaux d'hydroxyde sodium nécessaire pour C = 50 % ;

⟶ 50g de NaOH ⟶ 100 ml d'eau distillée. On prélève une masse de 50g qu'on introduit dans une fiole de 100mL puis on complète à l'eau distillée. C'est-à-dire que 500 ml d'eau ⟶ 250 g de NaOH   ③

<u>Annexe 3 :</u> Les équipements choisis les caractéristiques et les différents fournisseurs

Caractéristiques	Fournisseurs	images
<u>Nom : extrudeuse gonflage</u> Diamètre: 60 Diametrevis : 25 Entrefer: 0,6 Type de vis: standard 25D Laize à plat: 350mm. Taux de compression : 3 3 couches : A extérieure, B milieu, C extérieure Vitesse de rotation max: 160 Pression d'alarme: 300 **Débit maxi : 8kg /h**	**CFP (conseil formation plasturgie) , contact: 0472682828**	
<u>Nom : imprimeuse flexographique</u> Type : Gearless Couleurs : 8 , 10/8 Largeur d'impression (mm) : 670/870, 1070/1270/870 Format maximum (mm) : 600 – 800 Vitesse (m/min) : 300 - 400	**COMEXI, TEL : +34972477744**	

Nom : ondes ultrasonores Hielscher UIP 16000 technologie ultrasons L'appareil ultrasonique 200W UP200St Processeur à ultrasons industriel UIP16000 (16kW)	CFP (conseil formation plasturgie) ,contact: 0472682828	
Nom : centrifugeuse industrielle Nombre de disque : environ 70 Vitesse tournante : 4650t/mn Nombre de bec : 10 Puissance de moteur : 37 KW Capacité : ≤45m³ Poids : 1550 Kg	Flottweg tel : + 331 82726030	
Nom : étuve Marque: H954 étuve spécial sur mésure Volume de chambre : 2,25 m³ Température : 250 °C Données techniques : H954	France étuve	
Microscope 115V 20 watt bulb Large 10mm x 120mm the base is 5 x 7.5 inches	Fournisseur : entreprise locale	

Nom : Groupe électrogène Marque : BWCG ; Numéro type : Z.100 ; Tension évaluée : 400V/320V Type de rendement : C.A triphasé Garantie : 1 an ; Vitesse : 1500 Pm, silencieux ; Fréquence : 50Hz ;	Fournisseur : entreprise locale BERNABE AKWA DOUALA	
Nom : Bac à ordures Volume : 240 l Hauteur : 106,3cm Profondeur : 68,6cm Diamètre des roues : 30,4cm Resistance au vent (vide) : 80km/h Matière : PEHD	Fournisseur : entreprise locale BERNABE AKWA DOUALA	
Trieuse par gravité Technologie : par gravité	agriexpo.online	
Chariot transporteur Capacité : chargement (Kg) : 350 ; Volume (m3) : 60 ; Dimension : 1520mm (l) x 1135mm (h) x 720mm (w) ; Taille : L(100), l(80), H(105) ; Certification : iso 70004 N-0	Entreprise locale BERNABE AKWA DOUALA	

Cuve à eau (différents cuve à eau) Résistance à la température (40oC à -60oC) ; Capacité : 1000L d'eau --Dimension : 1200*1200*800	Entreprise locale BERNABE AKWA DOUALA	

Annex 4: Amounts to be paid for the creation of a limited liability company in Cameroon

Organizations concerned	Documents to be obtained	Amount to be paid
Registry of the Court of First Instance	Registration in the Trade and Personal Property Credit Register	51503 FCFA
CNPS	Certification for submission CNPS	8000 FCFA
	Certificate of non use of the salaried staff at the CNPS:	2500 FCFA
Tax Centre	Certificate of exemption from tax	Exemption for the first year
	Taxpayer Card	
Registration lease contract	10% of the annual lease amount declared by a developer Tenant	
Property tax	0,11% of the declared value of the building, for an owner entrepreneur	

Source: Chamber of Commerce, Industry, Mines and Handicrafts of Cameroon, (CCIMA) 2015.

Appendix 5: 2D view of the company

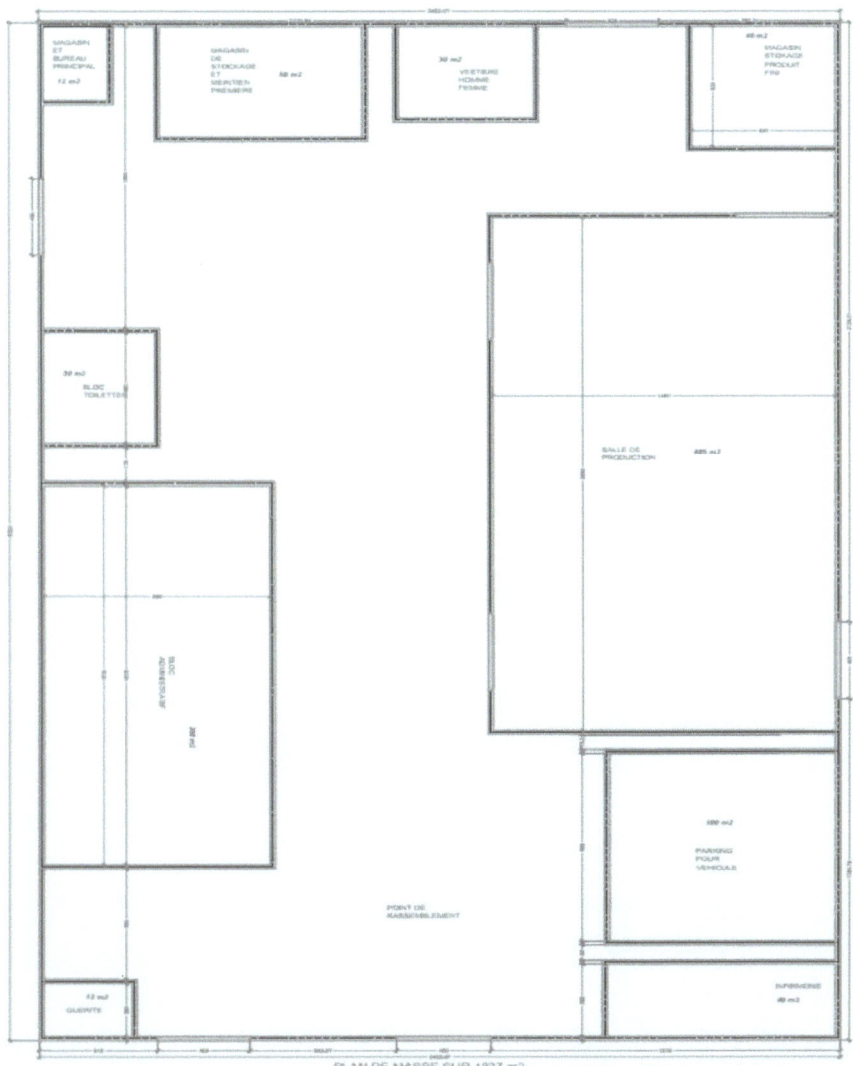

PLAN DE MASSE SUR 1837 m2

Printed by Books on Demand GmbH, Norderstedt / Germany